Routledge Revival.

Geopolitical Orientations, Regionalism and Security in the Indian Ocean

First published in 2004, this book is the inaugural volume of the Indian Ocean Research Group (IORG) and is based on a selection of papers presented at the IORG launch in Chandigarh in November 2002. The volume emphasizes the complexity and historical and contemporary geopolitical significance of the Indian Ocean Region (IOR). It also propagates the necessity for increased intra-regional cooperation, especially in terms of economic and environmental security, maritime boundaries, sea lane security and ocean management, in the spirit of open regionalism, in order to ensure a more secure IOR. In addition, the volume initiates an agenda for future social science policy-orientated research. The book should be of particular interest to policy-makers, business people and academics, as well as citizens of the IOR.

Dennis Rumley is currently Professor of Indian Ocean Studies and Distinguished Research Fellow at Curtin University, Western Australia, and was formerly Australia's focal point to the Indian Ocean Rim Academic Group (IORAG). He was appointed as Vice-Chair of the IORAG at the IORA meeting in Bangalore in November 2011 for a period of two years.

He has been an editorial board member of various international journals and has published 11 books and more than 130 scholarly papers on political geography and international relations, electoral geography, local government, federalism, Australia's regional relations, geopolitics, India-Australia relations and the Indian Ocean Region. He is currently Chairperson of the Indian Ocean Research Group (IORG) and is Foundation Chief Editor of its flagship journal, *Journal of the Indian Ocean Region*. IORG is one of two Observers to IORA.

Dennis Rumley edited a major study on Indian Ocean Security which was launched in Canberra by the Australian Deputy Foreign Minister in March 2013 - *The Indian Ocean Region: Security, Stability and Sustainability in the 21st Century*, Melbourne: Australia India Institute.

He has recently edited *The Political Economy of Indian Ocean Maritime Africa* (Pentagon Press: New Delhi), has co-edited *Indian Ocean Regionalism* (Routledge), with Timothy Doyle, and has had republished as Volume 15 in a new *Routledge Library Editions: Political Geography* series his co-edited book with Julian Minghi, *The Geography of Border Landscapes.*

Dr. Sanjay Chaturvedi is Professor of Political Science at the Centre for the Study of Geopolitics, Panjab University, India. He was awarded the Nehru Centenary British Fellowship, followed by Leverhulme Trust Research Grant, to pursue his post-doctoral research at University of Cambridge, England (1992-1995). He has also been a Third Cohort Fellow of India-China Institute at the New School, New York (2010-2012). He was the Co-Chair of Research Committee on Political and Cultural Geography (RC 15) of International Political Science Association (IPSA) from 2006 to 2012.

Founding Vice-Chairman of Indian Ocean Research Group (IORG Inc.), he is the Co-Editor of *Journal of the Indian Ocean Region* (Routledge) and the Regional Editor of *The Polar Journal* (Routledge). He also serves on the Advisory Board of *Geopolitics* (Routledge) and *Journal of Borderlands Studies* (Routledge).

Chaturvedi has served on the Indian delegations to the Antarctic Treaty Consultative Meetings (ATCMs) since 2007 and Track II Trilateral Dialogue on the Indian Ocean (TDIO) among India, Australia and Indonesia. His recent co-authored books include *Climate Terror: A Critical Geopolitics of Climate Change* (Palgrave Macmillan with Timothy Doyle), *Climate Change and the Bay of Bengal: Emerging Geographies of Hope and Fear* (Institute of South East Asian Studies, Singapore, with Vijay Sakhuja).

ARC ACKNOWLEDGEMENT

The editors would like to acknowledge the funding of the Australian Research Council Discovery Project DP120101166: "Building an Indian Ocean Region" for their generous assistance in the conduct and completion of this research, writing and publication 2012-2015.

Geopolitical Orientations, Regionalism and Security in the Indian Ocean

Edited by
Dennis Rumley & Sanjay Chaturvedi

Routledge
Taylor & Francis Group

First published in 2004
by South Asian Publishers

This edition first published in 2015 by Routledge
2 Park Square, Milton Park, Abingdon, Oxon, OX14 4RN
and by Routledge
711 Third Avenue, New York, NY 10017

Routledge is an imprint of the Taylor & Francis Group, an informa business

Publisher's Note
The publisher has gone to great lengths to ensure the quality of this reprint but points out that some imperfections in the original copies may be apparent.

Disclaimer
The publisher has made every effort to trace copyright holders and welcomes correspondence from those they have been unable to contact.

A Library of Congress record exists under LC control number: 2004328435

ISBN 13: 978-1-138-91666-1 (hbk)
ISBN 13: 978-1-315-68948-7 (ebk)
ISBN 13: 978-1-138-92316-4 (pbk)

Geopolitical Orientations, Regionalism and Security in the Indian Ocean

Edited by

DENNIS RUMLEY
University of Western Australia

SANJAY CHATURVEDI
Panjab University

SOUTH ASIAN PUBLISHERS PVT LTD.
50 Sidharth Enclave, PO Jangpura, New Delhi 110014

ISBN 81-7003-281-4

Typeset at
ANJALI COMPUTER TYPESETTING
50 Sidharth Enclave, PO Jangpura, New Delhi 110014
Phones: 26345315, 26345713
e.mail: sapub@ndb.vsnl.net.in

Published by
SOUTH ASIAN PUBLISHERS PVT LTD
50 Sidharth Enclave, PO Jangpura, New Delhi 110014
and printed at
RANGMAHAL
Pataudi House, Daryaganj, New Delhi 110002
Printed in India

CONTENTS

PREFACE

This book represents an important stage in the development of the Indian Ocean Research Group (IORG) since it is the Group's inaugural volume. The Indian Ocean Research Group (IORG) was launched at Panjab University in Chandigarh in November 2002. While the Group was created for a number of interrelated reasons, its central mission is to initiate a social science policy-oriented dialogue, in the true spirit of partnership among academics, policy-makers, business people, NGOs and communities towards realizing a shared, stable and prosperous future for all Indian Ocean inhabitants. Its principal objective is thus to act as a regional facilitator and coordinator of research on the nature and impact of the human uses of the Indian Ocean, with the overall aim of realizing regional peace, cooperation and ecologically sustainable development.

The IORG Principal Coordinators, Sanjay Chaturvedi, of Panjab University and Dennis Rumley, of the University of Western Australia, felt that there was an important need for such a Group and that previous regional attempts had been unsuccessful primarily because they were not built upon a strong bilateral base. The Coordinators felt that the development of a regional research network could be best facilitated by creating two regional 'pillars' in Chandigarh and in Perth, strengthening the relationship between the two pillars and then developing the network accordingly.

Thus, the IORG especially emphasizes the development of research networking, including distance education, among regional institutions of higher learning, the facilitation of dialogue among cultures and civilizations in the region, as well as acting as a resource base for data, information and expertise on the region. The IORG would endeavour to offer consultancy services to interested business groups and government agencies both to solve particular local and regional problems and to help facilitate intra-regional investment and trade.

The Group is currently structured around a series of interrelated research themes which are seen to be especially critical to the realization of IORG goals. These are: Energy Security, Environmental Security, Geopolitical Orientations of Indian Ocean states, Regional Cooperation, Sea Lanes, The Regional Role of External powers and Water Security. The present volume represents a selection of papers from some of the Group's researchers which were delivered at the Inaugural IORG Conference at Panjab University in November 2002.

We are especially pleased to be able to include within this volume the Inaugural Annual Indian Ocean Lecture which was delivered at the launch by Ms Penny Wensley AO, Australian High Commissioner to India. We very much appreciate Ms Wensley's support for the launch and the continued existence of the IORG.

The IORG aims to hold regular meetings on key integrative themes at different locations around the Indian Ocean Rim in collaboration with local host institutions. The second meeting, on the theme of 'Energy Security and the Indian Ocean', is to be held in Tehran in February 2004 in association with the Iranian Institute for Political and International Studies (IPIS). It is hoped that, over the next several years, IORG meetings will be held in Malaysia, Oman, Australia and South Africa, but the Group is always open to suggestions from regional institutions in any Indian Ocean state to collaborate in holding an international conference on any important regional issue.

Out of every meeting, the IORG hopes to publish at least one book which summarises its main research findings. Ideally, each book from the previous meeting will be launched at the subsequent meeting. Thus, at the second meeting in Tehran, it is intended to launch this volume. In the longer-term, it is hoped to launch a *Journal of Indian Ocean Studies* and IORG would be interested to hear from any likely publisher for this endeavour.

As the IORG network develops and matures, other activities will become increasingly important, including the sponsorship of low-level inter-state mutual cooperation and security dialogues, the development of intra-regional postgraduate programmes, distance education, summer schools and training programmes of various types. In terms of the latter, it is hoped that the IORG can identify a centrally-located tertiary institution which would be willing to function as a regional training node.

The IORG is always looking to expand its range of regional and extra-regional linkages and is always willing and interested to discuss research, training or consultancy proposals. To date, the IORG's principal problem has not been a regional lack of interest; on the contrary, there appears to be a great deal of regional support for this endeavour. Rather, the problem is one of seeking appropriate sponsorship, and, although some sponsorship has already been gratefully received, more is needed from regional business and regional governments.

some sponsorship has already been gratefully received, more is needed from regional business and regional governments.

In this regard, however, to date the IORG has received excellent support from Panjab University and from the University of Western Australia. In addition, the Australia-India Council, BHP-Billiton and the Indian Ministry of External Affairs have also been especially generous.

There are so many other individuals and groups who have been helpful in the finalization of this manuscript. In particular, Dennis Rumley would like to acknowledge the very generous Visiting Research Professorship provided by Kyoto University and for the supportive and creative working environment provided by all of his academic colleagues there. Kyoto University Vice-President, Akihiro Kinda, deserves special mention for his long-term support and vision. Sanjay Chaturvedi would like to thank all of the research scholars at the Centre for the Study of Geopolitics, Panjab University, and especially Miss Eva Saroch, for providing valuable research assistance.

Last, but not least, we would like to sincerely thank our publisher, Mr Vinod Kumar, South Asian Publishers, New Delhi, for bringing out this volume with tremendous enthusiasm, efficiency and good humour.

Perth, Australia   **Dennis Rumley**
Chandigarh, India   **Sanjay Chaturvedi**

CONTRIBUTORS

Sam Bateman, Associate Professor and Principal Research Fellow, University of Wollongong, Australia

Aparajita Biswas, Director of the Centre for African Studies, University of Mumbai, India

Christian Bouchard, Associate Professor, Department of Geography, Laurentian University, Sudbury, Canada

Sanjay Chaturvedi, Reader and Chair, Department of Political Science and Coordinator of the Centre for the Study of Geopolitics, Panjab University, Chandigarh, India

Radha D'Souza, Lecturer, School of Law, University of Waikato, New Zealand

Timothy Doyle, Associate Professor and Head, Department of Geography, University of Adelaide, Australia

Vivian Louis Forbes, Map Curator, University of Western Australia and Adjunct Associate Professor, Curtin University, Perth, Australia

Jean Houbert, Honorary Research Fellow, Department of Politics and International Relations, University of Aberdeen, Scotland

Graeme Hugo, Professor of Geography and Director of the National Centre for Social Applications of GIS, University of Adelaide, and Joint Director, Australian Centre for Population Research, ANU and University of Adelaide, Australia

Ashley Jackson, Research Fellow in the Humanities and Fellow of Mansfield College, Oxford University, England

Kenneth I McPherson, Mercator Professor, South Asia Institute, University of Heidelberg, Germany

P. V. Rao, Professor and Director of the Centre of Indian Ocean Studies, Osmania University, Hyderabad, India

Dennis Rumley, Associate Professor, School of Social and Cultural Studies, University of Western Australia, Perth

Vijay Sakhuja, Research Fellow, Institute for Defence Studies and Analyses, New Delhi, India

Penny Wensley AO, Australian High Commissioner to India

PROLOGUE

Charting Cooperation and Challenges: Australia and the Indian Ocean Region

Penny Wensley AO

The Indian Ocean is a vast geographical entity, and one with sometimes contested definitions. Those states with a presence or an interest in the Indian Ocean are diverse in every way – size, politics, economics, culture, religion. Further, the salient issues confronting these states when they look towards the Indian Ocean are likewise many and varied. This large region has a long, rich and complex history – whether of trade, cooperation and shared human creativity, or of conflict and suffering – and the repercussions of that history will be with us into this new century.

So, I have sought to focus my remarks a little by giving this paper the title of "Charting Cooperation and Challenges". In the maritime spirit of Indian Ocean regionalism, what I offer here is a chart, or a set of markers, as an aid to understanding the nature and the extent of the challenges confronting security and prosperity for Indian Ocean littoral states. It is very much an Australian perspective, which you may find useful to cross-reference against your own conceptual maps of the problems and opportunities for cooperation facing our countries.

Mature political relations among states, both bilaterally and multilaterally, are an essential overlay to cooperation in our region. However, in the chart that I propose, the practical parameters, the lines of longitude and latitude, are trade and security. It was trade – an extraordinary history of maritime trade – that made the Indian Ocean a region long centuries before the word "regionalism" was a fashionable term in diplomacy. Furthermore, it is security concerns that demand a special attention for all countries in, and with interests in, the Indian Ocean today.

Naturally, the definition of security I use here is a comprehensive one, including the host of transnational and sub-state challenges that have attained growing proportions in this era of globalization and interdependence. However, I will make particular mention of the two pressing global threats of the day, which have a direct bearing on the Indian Ocean region: the first is terrorism, given new and tragic prominence by the bombings in Bali in October 2002; and the second is the spread of weapons of mass destruction, as well as their possible use by states, such as Iraq, or by international terrorist groups.

Of course, the grid lines of economics and security are bound to intersect, and that is where some of the crucial policy opportunities and challenges for the region lie: in the protection of sealanes, the secure export of energy resources, the safeguarding of open economies from transnational crime and terrorism, and the contribution of economic growth to reducing disparity and improving human welfare and human security.

Having explored this framework of trade and security, as Australia's High Commissioner to India, I will also dwell on the foremost bilateral building block in Australia's engagement with the Indian Ocean region: the relationship between Australia and India, which has seen robust development in recent years.

For the purposes of this presentation, I am not proposing to apply a strict definition of the Indian Ocean region. I am assuming rather a broad definition of the region, since so many of its concerns cannot be considered in isolation from neighbouring regions, particularly the Asia-Pacific, where Australia has strong and abiding security and economic interests, but also the Persian Gulf, and, on environmental security issues, the Southern Ocean.

Australian Trade with Indian Ocean States

In the popular imagination, the Indian Ocean has enduringly been associated with commerce. It is a connection that goes back very many centuries, long before the colonial era. I am speaking economic linkages that are literally ancient. A refreshing and convincing school of historical interpretation in the past two decades has suggested the existence of a distinctive Indian Ocean world, based primarily on long-standing commercial links and trade routes. As Professor Ken McPherson has noted, in his valuable work of historical synthesis, *The Indian Ocean: A History of People and the Sea*, that the Indian Ocean has long been a great highway for people, goods and ideas. Unlike the Pacific or the Atlantic, the Indian Ocean had a history of regular shipping, trade and exploration, which dated well before the 16th century. Furthermore, this history was certainly not only one of European powers: merchants, sailors and fishermen from some of the world's earliest civilizations traversed its waters. The Indian Ocean has not merely served the interests of commerce, but has contributed to all of the positive impacts for human development that free trade brings with it, including the exchange of knowledge, culture, expertise and goodwill.

Some would argue that the integration of the Indian Ocean into a global trading economy, from the 18th century onwards, served to diminish this original Indian Ocean regionalism. Nonetheless, it is clear that the positive influence of the past has lived on. Indeed, despite the flux of history and the impacts of imperialism, the post-colonial era, the world wars and the Cold War, it is fair to say today that the Indian Ocean as a major maritime highway has never before known such traffic. By some estimates, it carries a third of the world's bulk cargoes and an even larger proportion of the world's oil, including the economic lifeline from the Persian Gulf to East Asia.

In line with the Indian Ocean's enduring commercial importance, it is to be expected that Australia has significant bilateral trading relations with so many countries in the Indian Ocean region. In aggregate terms, our total exports with the member states of the Indian Ocean Rim Association for Regional Cooperation (IOR-ARC) was almost A\$21 billion in 2001; our total imports were a little over A\$17 billion.

The nature of that trade becomes clearer through a few bilateral examples. I will mention four: India, South Africa, Iran and Sri Lanka.

Trade and investment between Australia and India has reached a new high. In the 2001-02 financial year, India became Australia's 13th largest export market. Two-way trade reached \$A3.3 billion, a 47 percent increase over two years. While bilateral trade is balanced in Australia's favour, its expansion has been in both directions. Along with exports, Australian investment is playing a key role in Indian economic growth. That investment is now \$A1 billion, much of it in fast-growing service sectors including IT, education, health, tourism, and leisure, media and finance, as well as mining and niche manufacturing. Increasingly, Australia is supplying industrial, agricultural and other products and services required by Indian manufacturers and consumers. For India, investment in Australia may remain quite small, but much of it is in IT and the important energy and minerals sector.

South Africa is by far Australia's largest and most dynamic market in Africa. It is our 23rd largest trading partner, with total trade valued at \$2.1 billion in 2001. Merchandise exports to South Africa were approximately \$1.3 billion in 2000-01, mainly in meat, crude petroleum, coal and aluminium. Wheat and alumina between them accounted for another $646 million of our exports. Motor vehicle exports to South Africa recovered to $39 million in 2001, from $15 million in 2000. Overall, our exports to South Africa have increased by 42 per cent over the last two years, with growth occurring particularly in ETMs, such as mining equipment, engineering equipment, electrical machinery, motor vehicles, medicaments and information technology items.

Australia's exports to Iran in 2001-02 were valued at A$784 million. Historically, Iran has been one of Australia's largest individual markets for wheat, which continues to dominate Australian exports to Iran. Australia-Iran trade relations are set to move to a new plane, however, following the successful visit to Iran two months ago by our Trade Minister and a large business delegation. Key outcomes of the visit included confirmation that Australia would provide a further 530,000 tonnes of wheat to Iran this year and the early conclusion of a technical services agreement by BHP-Billiton and PetroIran. Australia's Commonwealth Scientific and Industrial Research Organization (CSIRO) also announced its participation in the establishment of a new A$145 million Iranian Minerals Research Centre.

The dollar value of two-way merchandise trade between Australia and Sri Lanka has more than doubled since 1995 to A$429 million in 2001. Australia is the second largest foreign investor in Sri Lanka, and the total stock (level) of Australian investment in Sri Lanka is A$400 million. Pacific Dunlop's Ansell Lanka rubber products plant in Biyagama is the largest foreign investment in an industrial plant in Sri Lanka, and the largest single industrial enterprise in the country.

So, as these examples from the Australian experience suggest, a primary driver in the stable integration of the Indian Ocean region is bilateral trade, pure and simple. With progressive liberalization and the reduction of trade and investment barriers throughout the region, the great economic potential of so many countries around the Indian Ocean rim will increasingly be brought together to common benefit. Furthermore, in many cases, with the barriers down, the modern trade winds of market forces will ensure that the right goods, services and investments find the right markets, wherever they may be across this wide ocean. The ensuing contributions to economic growth, in conjunction with suitable governance, can help to reduce the wealth disparities that divide parts of the region.

Australia welcomes efforts to pursue freer trade in the region at all levels: whether bilateral free-trade arrangements, movement towards economic integration or trade in subregional contexts, such as the South Asian Association for Regional Cooperation (SAARC). Every step towards free trade in each of these settings has flow on benefits for us all, as competitive trading states. For instance, in 1998, Sri Lanka signed a free trade agreement (FTA) with India, the first stage of which came into force in February 2000. This has, in turn, provided opportunities for Australian companies to use Sri Lanka as a base for access to the Indian market.

Though it might seem to be stating the geographically obvious, I should add that trade with the Indian Ocean region is of particular importance to Western Australia. The point lies in the sheer magnitude of this business. In the 2000-01 financial year, West Australian companies exported merchandize worth around A\$5.2 billion to member states of the Indian Ocean Rim Association for Regional Cooperation. This was about a quarter of Australia's total exports to IOR-ARC countries for the same period.

One other characteristic of the trade and investment relationship Australia has and is pursuing with countries in the region, is the importance of the energy sector, and in that light I am pleased to note the participation tomorrow afternoon in the conference by the President of BHP-Billiton India, Mr David Twine. Energy is an important component of existing Australian trade with the region: in aggregate terms, petroleum was one of our major exports to the IOR-ARC countries in 2001 (worth A\$1.85 billion) and coal was another (A\$1.2 billion). In addition, our potential for exporting energy to the region is huge. Australia is also one of the world's largest producers and exporters of natural gas, and, each year provides around 7.5 million tonnes of liquefied natural gas or LNG to countries in the Asia-Pacific region.

Natural gas consumption is projected to rise strongly in most parts of the world; it is the fuel of choice for this century – a technically, economically and environmentally sensible solution. Australia's world class resources of natural gas mean it is ideally placed to be the LNG supplier of choice throughout the region. Proximity to emerging LNG markets such as India gives Australia a key advantage. The confidence and low risks in doing business with Australia weighed heavily in China's recent decision to sign a A\$25 billion contract for the supply of 3.2 million tonnes of Australian LNG a year for 25 years, starting in 2005. It is widely expected that India, too, will become a major buyer of LNG, into the coming decades. The Middle East may be a natural supplier to western India, which would have its own positive implications for regional cooperation. At the same time, however, Australia is strategically located to supply LNG to eastern India.

IOR-ARC and Australia

One recent and significant manifestation of Indian Ocean regionalism in trade has been the inter-governmental process that led to the formation of the Indian Ocean Rim Association for Regional Cooperation (IOR-ARC).

Launched in Mauritius in March 1997, the Indian Ocean Rim Association for Regional Cooperation (IOR-ARC) is a group of nineteen Indian Ocean states with the primary aim of facilitating trade and investment. The Association includes working groups that provide input from business and academia, with a view to reducing the costs of doing business in the region.

IOR-ARC is important to Australia's engagement with countries in the Indian Ocean Rim. Australia believes IOR-ARC can play a useful part in promoting closer economic engagement between itself and member countries in this region. In the shorter-term, IOR-ARC can engender a better understanding of the importance of economic cooperation and trade liberalization among member states. In the medium-term, the Association's focus on trade and investment facilitation should provide direct benefits to business.

Australia has played a key role in efforts to make IOR-ARC's work programme more results-oriented and focused on issues relevant to business. As a member of the high-level task force considering the future directions of IOR–ARC and the operations of its Coordinating Secretariat, Australia proposed and received solid support for a number of important reforms to the Association. These included recommendations on the submission of detailed project proposal documentation, and the evaluation of proposals for compatibility with the Association's core objectives. Australia took the lead role in drafting the final report, which was recommended to IOR-ARC ministers for adoption by the Council of Senior Officials at their last meeting in Oman in May 2002.

The Australian Department of Foreign Affairs and Trade has funded the research and publication of a number of useful reports on trade and investment issues in the Indian Ocean Rim. Studies on tariff structures, trade and investment regulations, and agriculture are consistent with our broader objectives, including in the WTO and APEC.

At the IOR-ARC Council of Ministers Meeting in Oman in April 2001, Australia presented the findings of a survey of Australian business, entitled *Enhancing Trade and Investment in the Indian Ocean Rim: An Australian Business Perspective*, which identified barriers to trade and investment within the Indian Ocean Rim and made recommendations which could produce direct benefits to business. At the Committee of Senior Officials Meeting in Oman in May 2002, Australia tabled a further three reports. These included the Murdoch University study on foreign direct investment (FDI) among IOR-ARC members and a *Guide for*

Business, which summarizes the key issues and findings covered in the FDI report, and is a source of relevant, tailored information for businesses active in IOR-ARC countries. The third report was Professor Kym Anderson's study, *Agricultural Trade Liberalization: Implications for IOR-ARC Countries,* published in July 2002.

These studies are consistent with the priority Australia places on improving transparency in trade and investment facilitation in IOR-ARC countries. They build on earlier work which sought to identify key impediments to trade faced by businesses in the region.

In particular, some members of IOR-ARC have identified the lack of foreign direct investment (FDI) as a limiting factor in their development, and the Australian study on FDI has provided them with a range of measures to encourage investment. The study on Agricultural Trade Liberalization serves to dispel some of the misinformation in the region regarding this issue and to encourage countries to pursue their trade priorities in the next WTO round.

At the same time as commending IOR-ARC for what it is, it is worth bearing in mind what this organization is not. Misperceptions or false expectations do not help in the development of regional cooperation. Sometimes, they can over-burden an institution or distract it from its primary task. IOR-ARC, as it exists, is not a vehicle for strategic dialogue or security cooperation, or for political cooperation beyond the domain of trade and investment. It is not regional security architecture, nor is it an organization in search of rapid expansion of its agenda. Rather, to the extent that IOR-ARC can meet the important trade facilitation goals already set for it, it will contribute over time to the region's prosperity, interconnectedness and stability – precursors for regional security.

From Trade to Security: The Maritime Dimension

A healthy web of trading relationships is central to the prosperity of the Indian Ocean region, and also to its stability. From Australia's perspective, a dependence upon trade for welfare and stability is not only positive, it is inevitable. Globalization has meant that Australia's economy, like all economies in the region, is more integrated with other states. Furthermore, an overwhelming proportion of this trade is carried by ship. In Australia's case, no less than 99.9 percent of this trade by bulk, and 71.7 percent by value, is carried by ship, making Australia the world's fifth largest user of shipping. Our access to world markets relies on merchant ships enjoying the freedom of navigation along trade routes. You can see then why the safety of what are technically known as Sea Lanes of Communication, including in the Indian Ocean, is of such importance to Australia.

Seaborne trade is vital to the economic growth and stability of the Indian Ocean region, just as to the Asia-Pacific, Australia's neighbourhood to the north and to the east. Much of this trade must pass through focal points such as the Malacca Straits. In the Indian Ocean, as in the Asia-Pacific, there is growing dependence on imported energy supplies, transported by sea, to sustain economic growth. Disruption to the freedom of navigation of maritime traffic – whether through terrorism, piracy or the impact of regional conflicts – would have serious consequences for regional economies.

The message here is that continued growth in seaborne trade is inevitable and welcome, but that our collective dependence on such trade, and the intrinsically exposed nature of merchant shipping on the open seas, introduces an element of vulnerability. Just as democracies and economically-advanced societies are especially vulnerable to acts of terrorism and other transnational threats, a region of trading states faces its own complex mix of security challenges.

Before I proceed to outline some of the key security challenges that face us all in the changing Indian Ocean environment, I thought it would be useful to give a sense of Australia's maritime perspective on the region.

The only continent that is a state, the only state that is a continent, Australia has logically moved towards a defence strategy that embraces the maritime. Australia is a maritime country. It is there in the first lines of the national anthem: "Our land is girt by sea". Girt may be a rather archaic word, but, interestingly, it has two meanings that both illuminate how Australia sees itself; indeed, what kind of country Australia is. Girt means encircled or surrounded, but it is also a nautical term meaning tightly moored or held in place. In other words, Australia as a country is, in large part, defined by the sea. I have already noted the enormous proportion of Australian trade that is seaborne. However, the sea – and this includes Australia's western seaboard – holds other profound economic value for Australia. Australia's 200 nautical mile Exclusive Economic Zone (EEZ), contains valuable fishing stocks and vast mineral and energy reserves. The Australian oil and gas industry is worth A$8 billion a year; most of these reserves are either offshore or transported in ships. Commercial fishing is our fifth largest primary industry, and 81 percent of this catch is exported. With island territories extending from the tropics to the Southern Ocean, Australia's maritime security responsibilities range well beyond the immediate shores of the mainland, covering 16 million square kilometres. In total, the Australian Navy patrols a zone of almost 10 percent of the world's total surface area.

The primary objective of Australia's defence policy is, of course, to prevent or defeat attacks upon Australia. Our geography and our reliance on the sea for trade makes it not surprising that the Australian Government, in its *2000 Defence White Paper*, adopted a maritime strategy as a cornerstone of the country's defence.

The growing importance of the sea in regional security thinking is reflected not only in Australian planning. Some other states in the Indian Ocean region, as in the Asia-Pacific, are developing impressive naval capabilities to support their national goals. Sophisticated weaponry and platforms are entering service in a number of regional navies.

Australia seeks to use its engagement with the navies of our countries to promote regional stability – a goal that holds true for the Indian Ocean as it does for the Asia-Pacific. Opportunities exist for international exercises and training with the armed forces of other states to develop professional skills while increasing mutual understanding. At the same time, the Australian Defence Force has shown itself ready to participate in important operations, including on the margins of the broader Indian Ocean region, to promote or restore peace and stability, typically under UN auspices. In recent years, the tempo and complexity of such operations has increased, and this trend looks likely to continue.

Security Challenges

So far, I have explored the central importance of trade, including in the energy sector, to the region's prosperity and peace, but have touched also upon Australia's recognition, as a maritime and trading state, of the vulnerabilities faced by a trading, maritime region. I want now to present a wider survey of some of the pressing security problems facing the region. I will do so with particular reference to the measures being pursued by Australia to help address these challenges. Many of these measures rely in large part on consultation and cooperation with regional partners. I will also refer in places to the Asia-Pacific experience in dealing with some of these concerns, bearing in mind that Australia is among a number of countries with substantial interests in the mutually-reinforcing security of both regions.

Terrorism, and the spread of weapons of mass destruction are the two critical dangers that demand the immediate attention of the Indian Ocean region and the wider international community. However, there are many other challenges – transnational in nature – that have the potential to become serious threats to our collective security, security in the comprehensive sense, if they are not addressed resolutely.

Of course, the emergence of these transnational threats to global and regional security does not make the long-standing security challenges we face in the Indian Ocean and Asia-Pacific regions go away. One fundamental objective for us all, of course, is to ensure stable and constructive relations among all states in the region. Peaceful relations among the larger powers are especially important in this regard. Australia, like many other Indian Ocean states, has been concerned at the state of relations between India and Pakistan. Australia was alarmed by the military build-ups earlier this year, and we have welcomed the movement towards de-escalation in the past few months, as well as the holding of elections in Jammu and Kashmir and in Pakistan. My Mission in New Delhi took an interest in following the conduct and course of the Jammu and Kashmir election: we welcomed it as a democratic process, and condemned the attempts to disrupt it through terrorism. Australia is encouraging an early resumption of dialogue between India and Pakistan to resolve all outstanding issues, including Kashmir. As demonstrated by the heightened tensions earlier this year – including what many saw as the real risk of war – the nuclear weapons factor in the region has not served as an influence for stability.

So, the rise of transnational threats does not mean that the traditional security concerns of peace and stability have gone away. Nor does it mean that the alliances and friendships we have counted on for decades are no longer relevant – quite the contrary.

The region has, in fact, had some history of efforts towards cooperative security. In 1971, Sri Lanka proposed an Indian Ocean Zone of Peace. In 1972, the UN formed the Ad Hoc Committee on the Indian Ocean, to prepare the way for a Conference to implement this proposal. Because of an inability to resolve fundamental disagreements, the Conference was never held. Like some other regional security initiatives with what seemed a laudable aim of reducing superpower rivalry, this one failed partly because it did not take account of a view that some regional states firmly held: that some external interest in the region can be a stabilizing factor.

In particular, Australia continues to see the United States naval presence in the Asia-Pacific and the Indian Ocean regions as a force for stability. Those who would tend automatically to disparage such a presence would do well to consider its continuing potential for upholding international stability in this new era of uncertainty. They might, for instance, look towards the cooperative model that India this year has quietly pursued: of patrolling the vital Malacca Straits in broad cooperation with US naval vessels.

WMD and Iraq

It would be impossible and wrong at the moment to talk about security threats in this region without beginning with Australia's perspectives on a very immediate and pressing danger: the proliferation of weapons of mass destruction, and in particular the case of Iraq. Although the Indian Ocean region remains somewhat undefined in a geopolitical sense, and Iraq may fall outside of it, instability in the Persian Gulf and the proliferation of weapons of mass destruction are justifiably sources of grave concern to a large number of states within this region.

The global terrorism threat has given new urgency to the international community's pursuit of disarmament and non-proliferation goals, and to our commitment to work together to prevent the spread of WMD – both to non-state and state actors. We have to be alive to possible links between terrorists, weapons of mass destruction, and certain states we know that are responsible for their proliferation. For over a decade Iraq has flouted legally-binding obligations to disclose and eradicate its weapons of mass destruction programmes. It has defied UN resolutions, UN inspections and UN sanctions.

There is no question as to Iraq's past use of chemical weapons, including against its own people. Furthermore, we know Iraq is actively pursuing chemical and biological weapons programmes, and has endeavoured to produce nuclear weapons. If Iraq's pursuit of these weapons continues, in a few years, we may be asking ourselves why we failed to act decisively when we still could. The international community has now drawn the line, with last week's Security Council Resolution and Saddam must disarm – there can be no more prevaricating, no more conditions, no more undermining of the UN's authority. We therefore support a tough new Security Council inspection regime for renewed, unconditional and unfettered access to suspected weapons of mass destruction sites. We hope for a peaceful resolution to this situation - the onus is now squarely on Iraq to allow this to happen.

Terrorism

The shadow of terrorism over some states and societies in the Indian Ocean region is a long one; it has been present for decades. For others, its danger has become especially stark in the past year. For Australians, I speak in particular of the horrific outrage perpetrated in Bali last month, which claimed so many Australians as well as Indonesians and other nationalities among its 180 or more victims, and before that, the carnage of 11 September 2001 last year. To Australia, the 12 October attack in Bali confirmed what we had long suspected and feared: the international terrorist threat extends to Australia's immediate neighbourhood, as demonstrated also by the arrest, in December 2001, of Jemaah Islamiyah terrorists planning to bomb Western diplomatic missions in Singapore. Australians are troubled about the reality of terrorism at our doorstop, and the possibility of it moving closer still. At the same time, we do not discount or ignore the calculated terrorist violence that has been inflicted upon innocent people elsewhere in the region, including in this country. Australia condemns unreservedly acts of terrorism committed in this country, and has reiterated that any support for terrorism within India from entities beyond its borders must cease.

We also have to be conscious of the troubling adaptability of terrorist groups and tactics, and the possibility that the nature of tomorrow's terrorism will be different yet again. In the Indian Ocean region, and other important global shipping routes, terrorism has the proven potential to take on new forms that we will need to find new and cooperative ways of addressing. I speak, for instance, of maritime terrorism, such as attacks on naval vessels and merchant shipping. We have recently witnessed an attack on a French oil tanker off the coast of Yemen, on the edge of the region, and there is, of course, the example also of the attack on the USS Cole. The potential for maritime terrorism warrants close attention from all states with interests in the region. In this regard, it is a positive development that India and the United States have submitted a proposal to the ASEAN Regional Forum (ARF) to host in India shortly a training workshop on facing maritime security challenges, including the maritime dimensions of terrorism and transnational crime. This workshop will seek to identify not only the problems but also the range of political, diplomatic and military means that may need to be deployed in response.

Given the transboundary reach of today's terrorist movements and tactics, the Australian Government is convinced that the problem needs to be tackled comprehensively by the international community: an inter-related package of actions taken domestically, bilaterally, regionally and in the global multilateral context.

Within Australia, we have strengthened domestic legislation to proscribe terrorist organizations and given agencies the powers they need to hunt down terrorist and sever their sources of finance. A number of the most dangerous terrorist groups active in countries in the Indian Ocean region have been proscribed in Australian laws against the financing of terrorism.

Australia is committed to international action against terrorism. Following 11 September, Australia, for the first time, invoked its ANZUS security alliance with the United States, and has contributed substantially to coalition anti-terrorist operations. An Australian Special Air Service detachment has been conducting operations in Afghanistan since November 2001. A three-ship Australian Navy task group is deployed at the edge of this region, currently undertaking maritime interception duties in the Gulf. Australian refuelling aircraft are providing support for coalition forces. At the same time, Australia – like India – has committed resources to rebuild the damaged society of Afghanistan to help stop the return of terrorist politics.

Security Cooperation

Australia is playing its part through diplomatic channels to strengthen the architecture of international cooperation to make the defeat of terrorism permanent. Australia is party to nine important international legal instruments of international anti-terrorism cooperation; we have this year signed a tenth, and are preparing to accede to an eleventh. In addition, Australia, like India, has played a leading role at the United Nations in efforts to finalize the proposed Comprehensive Convention Against Terrorism.

Australia has also turned its attention to regional action. Again referring to the Asia-Pacific experience, we have taken a lead role in developing ways in which that region's multilateral security body, the ASEAN Regional Forum, can combat terrorism. While I am not casting the ARF as an Indian Ocean institution, its achievements are in the Indian Ocean region's interests. The ARF contains states which have interests in the security of both regions – for instance, India, Australia, Indonesia, Malaysia and some other ASEAN states. Further, the terrorist problem which both regions face is indivisible.

The ARF's constructive focus has included intersessional mechanisms and training workshops to strengthen the regional capability against terrorism and high-level discussions on terrorist networks in South-East Asia. Australia is promoting in the Forum a proposal for regional cooperation in managing the consequences of terrorism. This envisages an agreement allowing countries to assist others hit by a terrorist attack, by providing search and rescue teams, medical specialists and facilities, engineers and other capabilities, as may be agreed.

Australia considers that strong bilateral cooperation between regional governments is critical in combating terrorism, especially in the exchange of information and intelligence to identify threats at the earliest possible stage. Money laundering is intrinsically connected to the viability of international terrorism. We have recently signed an agreement with Singapore to prevent money laundering, and shortly will jointly host, with Indonesia, a conference to combat terrorist financing and money laundering. On more direct matters of combating terrorism, Australia has moved promptly this year towards signing bilateral memoranda of understanding on counter-terrorism with Indonesia, Malaysia and Thailand. This is an idea which may also have potential in the Australia-India relationship. Certainly, we share a commitment to the fight against international terrorism, and I can say that Australia would support closer contact and dialogue with India on counter-terrorism.

Of course, while technical cooperation between states is important, it alone is not sufficient. Strengthening political will is equally important. As our Prime Minister said at APEC a few weeks ago, exhortation cannot substitute for the determination of individual governments, who know they have a terrorist problem within their borders, to do something about it.

Other Transnational Challenges

Terrorism is the most serious and confronting transnational threat to the security of this region. However, there are other transnational challenges – some with more than peripheral connections to terrorism – that warrant mention. I could refer to piracy, the illicit drugs trade, and the illegal small arms trade. I have already mentioned money laundering. There are also challenges related more to social development, which, nonetheless, could be described as valid security concerns, given their sheer scale and impact. Some parts of the wider region face problems of governance and lawlessness, and problems arising from developmental difficulties and economic disparities. Australia recognizes the continuing importance of policies designed to promote economic development, good governance, democratization and human rights, to provide conditions which underpin stability and security. Australia recognizes the need for a proactive approach to many of these concerns. For instance, Australia has paid particular attention to combatting the HIV/AIDS pandemic within the Indian Ocean region – a problem with which we have long experience and which we confronted domestically at a very early stage. Australia has a A$200 million global initiative in the fight against HIV-AIDS, and is committed to working with partner states, including India.

However, for detailed scrutiny, I will limit myself to two issues on which Australia's attention is closely fixed: the problems posed by illegal migration and people smuggling, as well as the environmental security challenge of illegal fishing.

A regional transboundary challenge which Australia has faced squarely is that of people smuggling and illegal immigration. People smuggling has become a significant source of funding for criminal organizations. People smuggling and people trafficking operate in close proximity with organizations engaged in the trafficking of narcotics, money laundering, arms smuggling, document fraud and the bribery/corruption of officials. There are also potential links to terrorism as a range of illicit services provided by these groups have applications that can assist in terrorist activities -– for instance, document fraud and clandestine cross-border movements.

Actions taken in combating particular forms of transnational crime may have flow on effects to other transnational crimes. It is widely recognized that a regional approach is required to effectively combat people smuggling, trafficking and related transnational crime.

Australia has adopted a comprehensive whole-of-government strategy to deal with people smuggling and to stem the illegal flow of people to Australia, with a view to ensuring it retains the capacity to control national borders and determine who enters the country, while, at the same time, taking careful account of our international human rights obligations. Australia is cooperating with a number of countries in the interception of illegal migrants, criminalization of people smuggling and prosecution of known people smugglers. Australia is also assisting other countries, including in this region, to strengthen their border control systems.

In keeping with the search for regional solutions to the problems of people smuggling and illegal migration, Australia and Indonesia co-hosted the Bali Conference on People Smuggling, Trafficking in Persons and Transnational Crime in February 2002. Australia is working with regional countries, the IOM and UNHCR to follow up recommendations of that Ministerial Conference.

India was represented at the first Bali Conference by its Ambassador to Indonesia, and has remained interested in the ongoing activities of the Bali process. India is also engaged regionally through the inter-governmental Asia Pacific Consultations on refugees, displaced persons and migrants. It faces its own challenges from illegal immigration. India has porous borders with several neighbouring states. Official estimates show that, for instance, a large number of Bangladeshis reside illegally in India. India is seeking to negotiate a returns agreement with the Bangladesh Government. India is also a transit country for Pakistani, Sri Lankan, Nepali and Middle Eastern nationals en route to the UK, US, Canada, Europe and Australia. However, I would note that India is not a large source country for illegal immigration to Australia. Australia is working with India in providing practical assistance through document detection training, to help reduce the problem of document fraud, which is such a major component of illegal immigration.

Illegal Fishing: Environmental Security

In the comprehensive definition of security that is applicable to the Indian Ocean today, we must not overlook the very fundamental issue of environmental security. As a maritime state, Australia places great importance on sustainable environmental management and economic development of ocean resources. We are committed to actively cooperating at the regional and global level to ensure both living and non-living resources are managed carefully and sustainably. We have a shared responsibility of ensuring the long-term health of the world's 7 oceans.

Australia has emerged as a world leader in the management, conservation and sustainable use of the marine environment and marine living resources and enjoys a strong international reputation for leadership in encouraging responsible ocean management. Australia is an active participant in the Global Programme of Action for the Protection of the Marine Environment from Land-Based Activities. Australia's Oceans Policy, launched in December 1998, states: *"Australia must participate internationally in bilateral and multilateral arrangements to establish and implement international regimes that are effective in identifying and addressing issues in transboundary management . . . Australia should provide leadership regionally and internationally in the management of our oceans, recognising the possibility that national activities may have effects on the marine jurisdictions of neighbouring countries."*

Sadly, despite the positive intentions of a considerable number of countries in the region and beyond, environmental security in the Indian Ocean and neighbouring waters is a battle very far from won. One issue Australia is deeply concerned about is illegal, unreported and unregulated fishing, otherwise known as IUU fishing. Australia's concerns stem not only from the immediate IUU fishing activity within the Australian EEZ, but also from the impact IUU fishing is having on global fish stocks. One particular example is the alarming rate at which the illicit plunder of stocks of the Patagonian Toothfish in the Southern Ocean that has been growing in recent years. I say the Southern Ocean, but, in effect, this type of activity is also an Indian Ocean issue. I am not talking here of legitimate fishing fleets being perhaps a little over-zealous, nor of the impact of traditional fishermen and their harvesting. We are looking here at nothing less that huge international criminal networks.

Their impact is not marginal: it amounts to taking Patagonian Toothfish at a magnitude higher than legitimate vessels are licensed to take. The effect of IUU fishing on Patagonian Toothfish stocks may mean that, within a few years, fishing of the stocks may then become unsustainable and the species itself may be under threat. In recent years, the Australian and French navies have attempted, with cooperation from legitimate fishing companies, to curb IUU fishing through patrolling, notably in the EEZ around the Heard and McDonald, Kurguelen and Crozet islands. But this is not enough.

Australia has taken active steps in trying to avert the potential extinction of species under threat from IUU fishing. In efforts aimed at combating IUU fishing, Australia has taken a lead role in multilateral organizations including the Convention on the Conservation of Antarctic

Marine Living Resources (CCAMLR) and the Convention on International Trade in Endangered Species (CITES).

Turning to the more general question of fisheries management in the Indian Ocean, Australia is committed to a range of existing regional and international arrangements and treaties, including the Indian Ocean Tuna Commission, the Convention for the Conservation of Southern Bluefin Tuna and the Asia-Pacific Fishery Commission. Australia supports the work of the Food and Agriculture Organization in this field, and is involved in current negotiations to establish a South West Indian Ocean Fisheries Commission (SWIOFC). Australia has consistently demonstrated its strong support for the exchange of technical expertise and other measures to improve international cooperation, including in statistics and data collection. We have impressive expertise on environmental security of the oceans, and we are sharing it to common benefit.

Australia cooperates with many countries and inter-governmental organizations on fisheries research, and is involved in technical and scientific working groups on many bodies, including the nascent South West Indian Ocean Fisheries Commission. Against this background, Australia recognizes as a priority the need to seek concrete outcomes in ocean resource measurement and management using existing mechanisms, before serious consideration is given to the development of new arrangements. While Australia was one of seven IOR-ARC countries to attend the first meeting of the Association's experts group on fisheries last year in Muscat, our delegation there made the point that any proposals for cooperative action should not overlap or conflict with the mandates of established bodies already managing fish stocks in the Indian Ocean.

We have come a long way, then, in describing the comprehensive security challenges facing the region, from the challenge of WMD proliferation in Iraq, to terrorism, and the raft of other transboundary challenges, including people smuggling and illegal fishing. I hope that, as well as giving a somewhat sobering account, I have made it clear where Australia stands in seeking practical solutions to these challenges. Despite the immediate dangers, there are nonetheless grounds for optimism about our region's security future, if we are watchful and focused on pragmatic, cooperative solutions.

The Australia-India Relationship

It would be absurd to conceive of a serious engagement with the Indian Ocean region without very close bilateral engagement with India, the country which, in so many ways, is the major regional power. It is not for solely sentimental reasons that the world knows this ocean as the Indian Ocean. I want to conclude by exploring some of the achievements and the prospects of the relationship between Australia and India. I am convinced that this is an important moment in relations between Australia and India – as indeed it is pivotal time in the shaping of India's place in the world.

The grid of economic opportunities and security challenges in the region that that I have outlined helps to illuminate the convergence of Australian and Indian interests in many areas. The challenge for our Governments is to pursue the potential partnership that comes with this convergence.

There are long-standing bonds between our countries: our nature as multicultural societies with democratic systems of government – indeed, we are among the most established democracies in the region; the legacy of our shared Commonwealth background; and other meaningful ties. It has been argued that there was a time in which our countries paid insufficient attention to the bilateral relationship. However, today the pace of Australia-India relations is picking up.

I have spoken already of the healthy shape of the trade and business relationship between our two countries. Moving to the political relationship, the high-level contact between us in recent years speaks for itself, with a substantial number of Ministerial visits in each direction since 2000, most recently the visit to India in April by the Australian Minister for Foreign Affairs, Mr Alexander Downer, for his second framework dialogue with your External Affairs Minister. These contacts are underpinned by frank and in-depth official dialogues, and are given further texture by private sector and people-to-people programmes and contacts, including those supported by the Australian Government through the Australia-India Council. On the human level, it is especially pleasing to note that Australia has, in recent years, moved to take its place as the third most popular overseas destination for Indian tertiary students.

Moving to the strategic domain, simple geography is one explanation for the convergent interests of our two countries. Descriptions of Australia and India as Indian Ocean neighbours are more than just symbolic when one sees the proximity of the Andaman Islands to Australia's Christmas Island. However, increasingly it is the shifting global and geopolitical realities which point toward convergence of interests and the logic of engagement.

Australia recognizes India is a major power. It is not surprising, then, that in Australia-India dialogue, subjects of security and strategic importance have led the way. In August last year, a fully-fledged bilateral strategic dialogue was held for the first time. The talks were open and constructive and demonstrated shared perspectives and common interests on a number of important issues, including in the Asia-Pacific and Indian Ocean regions. The delegates agreed that both countries were factors for stability in these regions. The agenda was wide-ranging – global and regional security issues, national security and defence policies, maritime security, and arms control and disarmament. Arrangements are now well on track for the next round of this dialogue to be held in Canberra next month. Meanwhile, the second round of a Track II dialogue with a similar agenda – the Australia-India Security Roundtable – was held in May 2002 in Sydney, supported by the Australia-India Council.

However, it is in the official arena where the next substantive steps can be expected. We look forward to our first direct military-military dialogue in December, due to be held back-to-back with the next strategic – or political-military – dialogue. We hope that this "mil-mil" dialogue, involving representatives of both countries' armed forces sharing views about their defence organizations and activities, will refine the strategic direction of the defence relationship and contribute to a practical prog- ramme of joint activities in the phase ahead.

As I noted earlier when I spoke of our memoranda of understanding on counter-terrorism with Malaysia, Thailand and Indonesia, there is a growing recognition between Australia and India, too, of the mutual benefits to be gained by a more formalized process of exchanging useful information. In identifying the security concerns that Australia and India share, and on which our perspectives are complementary, I would place particular emphasis on countering terrorism and other transnational threats to our democratic societies and our growing economies.

Given that the salience of transnational security challenges will continue to grow, and cooperative efforts to combat them must intensify, it is in Australia's and India's interests to build cooperation sooner rather than later. In the maritime dimension, we have an active programme of ship visits and participation in fleet reviews, but it is easy to see the potential to take the relationship further. I have spoken of the major importance of maritime issues for the security of Australia and its region, and I note India's demonstrated interest in practical engagement with others in this area. Against the background of growth in seaborne trade, growing energy needs in the region and the existence of transnational threats, it is clear that Australia and India have converging and have complementary interests in maritime security, especially in the north-east Indian Ocean.

Conclusion

To recapitulate, the Indian Ocean is a region of vast political, cultural, religious and economic diversity, which has historically been united, in a sense, by the movement of people, goods and ideas, between Asia, Africa and Australasia. It carries a huge proportion of the world's shipping and trade; it contains the great trading sea lanes linking Europe, the Gulf and

Asia, and has very considerable energy reserves. I have set out an Australian "chart" of economics and security in the region, highlighting the positives of trade and the challenges of terrorism and other dangers. Progress is being made, both in bilateral trade relations and in regional arrangements, towards further facilitating trade in the region, to common benefit. Yet this is also a region that has its share of security problems. The region's very virtues – of massive seaborne trade, of growing energy demand and supply, of complex interaction between peoples and states – make it vulnerable to transnational threats, including terrorism.

Australia has a natural and considerable engagement with this region in trade, and strong interests in Indian Ocean security, and, in recent years, it has consolidated some key political relationships in the region, notably with India. Given the salience of key security concerns, including terrorism, weapons of mass destruction, transnational criminal activity, and the possibility of disruption to vital ocean traffic, Australia is conscious of the need to enhance practical cooperation in the region in the broad security field, as well as in trade facilitation. Much of this cooperation is being pursued at a bilateral level; for instance, in the central example of growing India-Australia cooperation. It is not Australia's aim to create institutions or expand agendas for their own sake, but rather to focus on core issues and areas where existing arrangements or attempted remedies are not effective. Australia applies this commonsense logic equally to its engagement with the Asia-Pacific as to our other important ocean neighbourhood, the Indian Ocean region. A strong web of mature bilateral relationships will be essential to our success in building shared prosperity and stability and defeating transnational and other threats in the Indian Ocean region. For my part, I look forward to further consolidating our relationship with India as a centre-piece of this.

[Text of the Inaugural Indian Ocean Lecture, Panjab University, Chandigarh, 17 November 2002]

CHAPTER 1

Changing Geopolitical Orientations, Regional Cooperation and Security Concerns in the Indian Ocean

Dennis Rumley and Sanjay Chaturvedi

Introduction

The main purpose of this Chapter is to identify some of the central issues raised in this book and to introduce each of the contributions in relation to these aims within the three broadly-defined themes of changing geopolitical orientations, regional cooperation and security concerns. However, before doing this, a brief discussion is undertaken of the question of Indian Ocean regional definition.

The Indian Ocean as a Region

From an academic geographical perspective, one of the broadest definitions of the region is that it includes 47 littoral and land-locked states bordering on the Indian Ocean (IFIOR, 1995). However, if all ASEAN states are included, then this raises the total number of regional states to at least 50 (Figure 1.1). Such a definition results in a very large region of considerable ethnic, religious, economic, political and cultural diversity with little apparent commonality. However, it has been argued that, prior to European colonial contact, the economies of the Indian Ocean region comprised a self-conscious "world" (McPherson, 1993, 5). While there is presently a relatively low level of functional cohesion among regional states, the orientation to the Ocean creates a degree of common interest and forms the basis for a potentially greater degree of

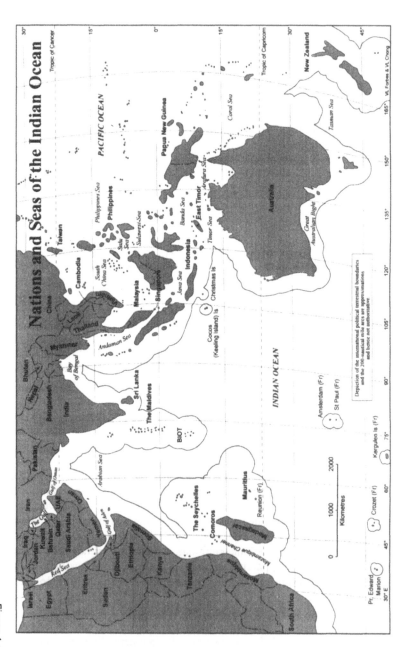

Figure 1.1

functional interaction in the future. Furthermore, apart from the colonial heritage, one critically important commonality of Indian Ocean states is that the majority are members of the developing world and few states possess high human development indices. Common developmental problems (Kerr, 1981; Appleyard and Ghosh, 1988) and their implications for national and regional security, defined in the broadest sense, can also provide a basis for increased South-South cooperation.

The academic and the practical policy definition of the Indian Ocean as a region is an inevitably contested issue as a result. The principal reason for this is fairly obvious since all regions are constructed, and, depending on the primary purpose of this construction – academic, administrative, economic, or, whatever – different regional definitions result. In short, the Indian Ocean, as a constructed region, will be defined by those who are involved in the construction process and whose collective goals will determine the composition of regional states. From the perspective of various types of regionalisms – economic, security and environmental – regional definition, of course, also depends on those states willing to join and remain with the group. This essentially 'inclusive-exclusive' debate runs through all global regionalisms, and the Indian Ocean region is no exception. For example, the Seychelles, which joined the Indian Ocean Association for Regional Cooperation (IOR-ARC) in 1999, decided to withdraw in 2003. Other states, including Pakistan, are currently excluded from IOR-ARC. Furthermore, in the case of the Asia-Pacific Economic Cooperation (APEC) grouping, contrary to some expectations, India was excluded from regional membership in 1997. What these examples clearly indicate is that regional construction is essentially a geopolitical exercise and that regional membership will be determined geopolitically. In the final analysis, the fundaments of regionalism and regional construction derive from the geopolitical orientations of member and non-member states in the context of global, regional and national imperatives. Furthermore, regions will expand or contract in direct response to the dynamics of these imperatives.

Changing Geopolitical Orientations in the Indian Ocean

It is not uncommon for Western commentators to underestimate the geopolitical importance of the Indian Ocean region. There are at least five interrelated reasons why this might be so. First, is that some observers see the region only as comprising the "Third" world, and thus, by implication, it is of lesser significance. Second, and following from this, some Western observers simply remain ignorant about the Indian Ocean and about the geopolitical orientation of regional states due to the existence of other ocean priorities, such as the Pacific or the Atlantic. Third, those commentators who underestimate the geopolitical importance of the Indian Ocean, to some degree, still exhibit a kind of 'colonial' perspective; as if the region had been frozen in time prior to the decolonization of most regional states. Fourth, ignorance can also be as a result of the fact that more needs to be written about the geopolitical importance of the Indian Ocean in general, and the geopolitical orientations of Indian Ocean states, in particular. The latter is especially problematical from the viewpoint of the limited number of available works on regional states written from an 'inside' perspective – for example, for the "Indian Ocean triangle" states, Australia (Rumley, 1999), India (Panikkar, 1955; Chaturvedi, 1998) and South Africa (Mills, 1998). A fifth point is that few regional state governments provide any explicit public discussion of their Indian Ocean orientations, with the notable exceptions of India's 'Look East' strategy and Australia's 'Look West' policy. How many of them are actually oriented towards the Indian Ocean it itself a debatable issue

As has been pointed out, however, the Indian Ocean possesses considerable geopolitical importance, if only because of its operation as a "highway" (McPherson, 1993, 5). Trade interactions among Europe, the Middle East and East Asia, for example, all rely on uninterrupted access through the Indian Ocean. Furthermore, as Christian Bouchard points out in Chapter 4, the Indian Ocean region comprises a number of sub-systems of varying geopolitical significance. He argues that it is only through the application of multi-scale analysis that the complexity of Indian Ocean geopolitics can be grasped. The Persian Gulf sub-system, which contains the greatest regional concentration of global oil reserves, also is a region of high internal fragmentation while possessing above average GNP per capita. Social, economic and political stability within these regional states and freedom of access through the Indian Ocean and contiguous seas by large oil consumers in the North is of inestimable importance to global economic stability.

The changing geopolitical significance of the Indian Ocean can be conveniently envisaged as comprising four principal stages – a pre-colonial phase, a colonial phase, a Cold War phase and the present post-Cold War 'globalized' phase. In Chapter 2, Ashley Jackson discusses the relevance of British colonial interests in the Indian Ocean from the eighteenth century until the twentieth century, focusing on the region's strategic significance. While the Indian Ocean *Pax Britannica*, which was established after the Revolutionary and Napoleonic Wars, was shattered in 1941 by Japan, the region continued to be one of British hegemony until the 1960s, albeit as part of a reconfigured global order underwritten by United States' power.

To a considerable degree, the geopolitical importance of Indian Ocean states increased considerably during the Cold War phase. Superpower rivalry, especially from the late 1960s, propelled the search for regional client states in strategic locations, and especially those which were reaching independence and had access to or proximity to important regional resources. At about the same time that the superpowers were 'invading' the British Lake, the British themselves decided to vacate it, and, by implication, Western strategic interests were left in the hands of the United States. To some extent, the historical legacy of the structure of Cold War linkages in the Indian Ocean region still inhibits regional cohesion.

There are yet others who would discount the prospect for a consideration of Indian Ocean geopolitics due to the impacts of the processes of globalization. However, as Jean Houbert argues in Chapter 3, the global system of states remains as important today as the transnational capitalist economy. Furthermore, he suggests that, as far as the Indian Ocean region is concerned, the configuration of land and sea still remains highly significant geopolitically. With decolonization on land, power at sea actually became more important.

In the post-Cold War period, Christian Bouchard in Chapter 4 argues that the Indian Ocean region has entered a new geopolitical era; what he refers to as the "Indianoceanic order". The structure and dynamics of this new "order" are seen to be articulated around five principal characteristics – regional heterogeneity, a system of Indian Ocean sub-regional sub-systems, the emergent IOR-ARC, a subordination to large foreign powers, and the geopolitical importance of the Ocean itself.

Regional Cooperation in the Indian Ocean

With the end of the Cold War, ideological regionalism eventually began to give way to economic regionalism, and this 'new regionalism' took less note of former ideological divides and incorporated states at different levels of economic development. In part, such a development was a response to difficulties with multilateral arrangements, but it was also a response to concerns over some of the negative impacts of globalization. Furthermore, states felt that there was a direct correlation between economic regionalism and national economic growth, and that, by being part of a regional arrangement meant that, not only would it ensure that they would not be 'left out', but that they might be able to influence the behaviour of other regional states in ways which would not be possible outside of regional arrangements. For states of the South, all of these reasons had particular resonance. However, the Indian Ocean was faced with a particular contextual problem because it did not possess a strong regionalism tradition. Furthermore, in many other cases around the world, regionalism was built on relatively strong bilateral linkages. In the Indian Ocean, however, relatively weak bilateral relations implied a relatively weak regionalism from the outset. On the one hand, the creation of the Indian Ocean Association for Regional Cooperation (IOR-ARC) in 1997, nonetheless, was a significant development. On the other hand, by dint of IOR-ARC membership, only 18 states around the Indian Ocean Rim currently explicitly possess some degree of pan-Indian Ocean regional commitment (Table 1.1).

Furthermore, as Ken McPherson argues in Chapter 5, there has been a reduced level of interest for IOR-ARC in recent years due mainly to its primary emphasis on the bureaucratic control of trade and investment issues. This has had two principal outcomes; first, it has meant that other 'second track' groups have become essentially irrelevant to the operation of the regional organization, and, second, that many other important regional matters, ranging from HIV/AIDS to comprehensive security, are unable to be discussed. In short, for most Indian Ocean inhabitants, IOR-ARC is likely to have little relevance, unless and until regional state governments are prepared to refurbish IOR-ARC with a new vision and are willing to commit sufficient resources to ensure a positive and proactive commitment on the part of regional business and academic groups.

Table 1.1. Indian Ocean Association for Regional Cooperation (IOR-ARC) States.

Australia (1997)
Bangladesh (1999)
India (1997)
Indonesia (1997)
Iran (1999)
Kenya (1997)
Madagascar (1997)
Malaysia (1997)
Mauritius (1997)
Mozambique (1997)
Oman (1997)
Seychelles (joined 1999, withdrew 2003)
Singapore (1997)
Sri Lanka (1997)
South Africa (1997)
Tanzania (1997)
Thailand (1999)
United Arab Emirates (1999)
Yemen (1997)

Thus, as P. V. Rao suggests in Chapter 6, it seems that sub-regional groupings – ASEAN, GCC, BIMST- EC - rather than a pan-Indian Ocean regionalism, are more important in dealing with regional problems. Furthermore, the economic and political case is put for a potentially new arrangement – the Kunming Inititative – which could bring together the peripheries of Bangladesh, Burma, China and India within a framework of regional cooperation.

The role of South Africa in the Southern African Development Community (SADC) is discussed by Aparajita Biswas in Chapter 7. Southern African regional states are concerned about some of the potentially negative impacts of this prospect of South African hegemony. However, even within South Africa, regionalism is contested along three principal lines. First, there is the view that South Africa should pursue its own interests while being resistant to regional concerns. Second, one of the contested elements of regionalism theory is that successful regionalisms require a regional hegemon, and there is the view that South Africa should actively pursue this as a goal. Third, there are those who favour a more cooperative mutually-beneficial regionalism which might include some form of security regionalism.

Security Concerns in the Indian Ocean

From a global perspective, it has been suggested that, geopolitically, the Indian Ocean does not comprise a "true security system"; rather, it comprises a set of highly diverse sub-systems. In brief, the Indian Ocean has been characterized as being akin to an "arc of crisis", since it contains many poor, unstable states which are beset by bilateral disputes (Gordon, 1996). This view is similar to an earlier characterization by Brezhinski of the region as an "arc of crisis" through the Middle East into Africa (House, 1984, 6). Many regional states are as much concerned with the problems of internal stability as they are with regional issues or questions related to an external threat.

In terms of the most basic indicator of traditional security – defence expenditure – the Indian Ocean region contains two of the world's top ten spenders – Saudi Arabia (US$21 billion) and India (US$13) (Commonwealth of Australia, 2003a, 20). In addition, both Australia (US$7.5 billion) and Indonesia (US$6.2 billion) possess very substantial defence budgets, as do Iran (US$4.9 billion), Kuwait (US$3.3 billion), Malaysia (3.3 billion), Burma (US$2.8 billion), Oman (US$2.7 billion), United Arab Emirates (US$2.7 billion), Pakistan (US$2.5 billion) and South Africa (US$1.7 billion). As a proportion of GDP, it seems that few states have increased their defence expenditures, 1990-2000, however (Table 1.2).

Table 1.2. Some Conventional Security Indicators in the Indian Ocean Rim.

	Armed Forces (total force)	Militarization (force as % pop)	Arms Imports US $M 2001	Defence as % GDP 1990	Defence as % GDP 2000
India	1,303,000	0.1	1,064	2.7	2.4
Pakistan (L)	612,000	0.5	759	5.8	4.5*#
Iran	513,000	0.8	335	2.7	3.8#
Egypt	448,000	0.7	486	3.5	2.3
Burma	344,000	0.8	n.a	3.4	1.7*#
Thailand	301,000	0.5	162	2.2	1.6
Indonesia	297,000	0.1	38	1.3	1.1#
Saudi Arabia	202,000	1.0	143	2.8	11.6*
Bangladesh (L)	137,000	0.1	180	1.0	1.3
Sri Lanka	15,000	0.6	40	2.1	4.5*#
Malaysia	96,000	0.4	20	2.6	1.9#
Yemen (L)	66,000	0.4	33	8.5	5.2
UAE (H)	65,000	1.7	288	n. a.	n. a.
South Africa	63,000	0.2	17	3.8	1.5
Singapore (H)	60,000	2.0	141	4.8	4.8*#
Australia (H)	51,000	0.3	687	2.2	1.7
Oman	44,000	1.8	30	18.3	9.7*#
Tanzania (L)	34,000	0.1	n. a.	2.0	1.3
Kuwait (H)	15,000	0.6	34	48.5	8.2*#
Mozambique (L)	6,000	0.04	n. a	10.1	2.5
AVERAGE		0.6	262.2	7.3	3.8

All states have medium human development levels unless otherwise noted (H or L)

*military expenditure higher than on health

#military expenditure higher than on education

*# military expenditure higher than on health and education

Source: United Nations Development Report 2002

While the Indian Ocean region contains considerable variation by state in terms of the size of the armed forces, the level of militarization does not necessarily correspond. Thus, while India has the region's largest armed forces of more than 1.3 million, its militarization index is well below the regional average at 0.1. Kuwait, on the other hand, with armed forces of 15,000, has a much higher level of militarization (Table 1.2). However, the regional arms trade is significant and there are considerable regional arms imports, especially to India, Pakistan, Australia, Egypt, Iran and United Arab Emirates, which, to a considerable degree, reflect internal regional conflict and/or instability.

Unfortunately, given the developmental context of many Indian Ocean states, international as well as domestic realities ensure that often human development takes second place to military development. There are at least six regional states – Burma, Kuwait, Oman, Pakistan, Singapore and Sri Lanka – in which government military expenditure exceeds government expenditure on health and on education (Table 1.2).

In the world of the 21st century, however, what is becoming increasingly clear is that the traditional realist model of security – principally embodied in the view that threats emanate from one state, are aimed at another state, and are of a military nature – has increasingly come into question (Commonwealth of Australia, 2003b). Furthermore, the specific nature of these 'non-traditional' security threats – for example, environmental security threats, people smuggling, drug trafficking, piracy and terrorism – require 'non-traditional' responses by state security agencies (Dupont, 2003). At the Indian Ocean regional scale, these new security threats essentially require a new approach to regional security cooperation at various levels. In short, regional security issues can only be resolved regionally.

The post-Cold War period has seen an increasing importance of environmental security questions at the global (for example, Barnett, 2001; Manwaring et al, 2002), regional (for example, Dupont, 2001) and ocean (for example, Broadus and Vartanov, 1994) scales, although both the concept and approaches are contested (for example, Deudney and Matthew, 1999). Nonetheless, the last decade or so has seen the emergence of a considerable environmental security discourse replete with its own concepts – for example, environmental threat, green aid, environmental colonialism, environmental piracy, environmental terrorism, ecological attack, green war and environmental refugees (Rumley, 1999, 196). As Timothy Doyle shows in Chapter 8, since there has been little systematic work completed at the scale of the Indian Ocean on environmental security matters, there is a great need for a new research agenda. Such an agenda, which incorporates regional issues of land degradation, water, fisheries, climate change, nuclear waste, environmental refugees and the impacts of the regional urban explosion, requires an appropriate regional data base and problem-solving strategies. In Chapter 9, Radha D'Souza argues for the need to reconceptualize or 're-envision' transboundary water disputes as being essentially developmental, rather than being necessarily geopolitical conflicts. In particular, she argues that there is a need to critically evaluate the role played by different organs of the United Nations in terms of the ways in which they operate to reinforce dominant power structures and dominant discourses through which the world is perceived.

The overall question of international migration raises many questions concerning overpopulation and population policies as well as many other sensitive non-traditional security questions concerning unauthorized migration, refugees and humanitarian migration programmes. After a detailed analysis of India-Australia migration, in Chapter 10, Graeme Hugo raises many pertinent national and regional issues concerning the relationship between migration and demographic growth; migration and social cohesion; migration and economic growth; refugee and humanitarian migration; "undocumented migration"; immigration and regional development; immigration and poverty; temporary migration and the "immigration industry".

Non-traditional threats to security are also evident in and on the Indian Ocean itself, in addition to pollution of the Ocean and the need for a regional approach to Indian Ocean management. In Chapter 11, Vivian Louis Forbes sets the regional scene by amplifying the jurisdictional framework within which current maritime boundaries are determined. He then examines a number of Indian Ocean case studies and emphasizes the prospects for inter-state and regional cooperation in maritime affairs. Vijay Sakhuja is especially concerned with threats to sea-lanes in the Indian Ocean in Chapter 12. He examines both the external and internal factors which determine regional acts of piracy, terrorism and other illegal activities and argues convincingly for greater regional cooperation on these issues. Sam Bateman makes a similar plea in Chapter 13, while emphasizing the necessity for the freedom of navigation in the Indian Ocean due to the economic and strategic importance of ocean shipping. He argues that, while there is scope for increased regional maritime and naval cooperation to enhance freedom of ocean navigation, it is also important to ensure that cooperative activities become 'building blocks' for, rather than 'stumbling blocks' to, a more secure Indian Ocean.

References

Appleyard, R.T. and Ghosh, R.N., eds. (1988), *Indian Ocean Islands Development* (Canberra: National Centre for Development Studies, ANU).

Barnett, J. (2001), *The Meaning of Environmental Security: Environmental Politics and Policy in the New Security Era* (London: Zed Books).

Broadus, J.M. and Vartanov, R.V., eds. (1994), *The Oceans and Environmental Security: Shared US and Russian Perspectives* (Washington, DC: Island Press).

Chaturvedi, S. (1998), "Common security? Geopolitics, development, South Asia and the Indian Ocean", *Third World Quarterly*, Vol. 19, No. 4, pp. 701-724.

Commonwealth of Australia (2003a), Advancing the National Interest: Australia's Foreign and Trade Policy White Paper (Canberra: DFAT).

Commonwealth of Australia (2003b), *Australia's National Security: A Defence Update 2003* (Canberra: DOD).

Deudney, D.H. and Matthew, R.A. (1999), *Contested Grounds: Security and Conflict in the New Environmental Politics* (State University of New York).

Dupont, A. (2001), *East Asia Imperilled: Transnational Challenges to Security*, (Cambridge: Cambridge University Press).

Dupont, A. (2003), "Transformation or stagnation? Rethinking Australia's defence", *Australian Journal of International Affairs*, Vol. 57 (1), pp. 55-76.

Gordon, S. et al (1996), *Security and Security Building in the Indian Ocean Region* (Canberra: Strategic and Defence Studies Centre, ANU).

House, J.W. (1984), "War, peace and conflict resolution: towards an Indian Ocean model", *Transactions*, Institute of British Geographers, New Series, Vol. 9, pp. 3-21.

IFIOR (1995), "Current economic characteristics and economic linkages in the Indian Ocean", *Working Paper*, Number 2.

Kerr, A., ed. (1981), *The Indian Ocean Region: Resources and Development* (Nedlands: University of Western Australia Press).

Manwaring, M.G. et al, eds. (2002), *Environmental Security and Global Stability: Problems and Responses* (New York: Rowman and Littlefield).

Mills, G. (1998), "South Africa and security building", *Canberra Papers on Strategy and Defence*, No. 127, ANU, Canberra.

McPherson, K. (1993), *The Indian Ocean: A History of People and the Sea* (Delhi: Oxford University Press).

Panikkar, K.M. (1945), *India and the Indian Ocean: An Essay on the Influence of Sea Power on Indian History* (London: Allen and Unwin).

Rumley, D. (1999), *The Geopolitics of Australia's Regional Relations* (Dordrecht: Kluwer).

GEOPOLITICAL ORIENTATIONS

CHAPTER 2

The British Empire in the Indian Ocean

Ashley Jackson

Introduction

This chapter outlines British interests in the Indian Ocean from the eighteenth century until the twentieth century, with a focus on the region's strategic significance. Why is this relevant in a book focusing on contemporary geostrategic issues in the Indian Ocean? Firstly, the British Empire was the primary interface connecting the region to an increasingly globalized modern world. Secondly, it is historically important to understand how for a century and a half, a single European power maintained supremacy and pursued its interests in the Indian Ocean, and how, in the mid-twentieth century, its control was almost overturned by the naval power of Japan – a 'might-have-been' that would have profoundly altered the region's development. The Indian Ocean *Pax* Britannica established after the Revolutionary and Napoleonic Wars was shattered in 1941-42 when Japan threatened to extinguish British power in the East and strike across the Indian Ocean to the shores of Africa – perhaps linking with German forces attacking the British position in the Middle East and so cutting the Empire in two.

Finally, until the 1960s, the Persian Gulf and Indian Ocean continued to be regions of British hegemony, albeit in a reconfigured global order in which American power underwrote that of its Western allies during the Cold War. Even during the decolonization years, Britain retained the capacity and will to project military power throughout the region. Its presence was facilitated by the retention of some colonies, continued relations with other territories of traditional 'informal' interest (for example the Trucial States of the Gulf), and the granting of base rights by newly-independent countries (the Royal Navy remained in Ceylon until 1957, and Singapore until the 1970s). The 1960s witnessed numerous British military interventions in newly-independent states in the Indian Ocean region, such as Malaysia, Kuwait, Tanzania, and Mauritius. Even with Britain's withdrawal from east of Suez, announced by Prime Minister Harold Wilson in 1967, the British role lingered on, and its one remaining colony – British Indian Ocean Territory – became the centre-piece of America's military presence and capacity in the region. It remains to this day a strategic factor of the first order.

Therefore, historical background helps situate the chapters that follow, many of which deal with twenty-first century themes and strategic preoccupations in the Indian Ocean, that, to the historian of the British Empire, have a ring of familiarity about them. Links among India, Australasia, Persia, and Africa have preoccupied policy-makers in the region, indigenous and alien, for many centuries. Today's concerns about strategic access, piracy, regional communications, natural resources, energy links, and trade, mirror those that exercised British imperial policy-makers in an era of European colonial rule that should not be overlooked as a vital precursor to the modern age.

The first half of the chapter describes the development of British imperial interests in the Indian Ocean and the implications of the changing balance of world power for British control in the region. Though the focus is upon the British presence, it is acknowledged that pre-existing connections shaped the Indian Ocean world and that non-Europeans had been the driving force in forging them. Even during the period of British imperial preponderance, Indian migrants continued to be key actors in inter-regional links, and India under the Raj was a sub-imperial force autonomous of London whose weight was felt from the Swahili coast to the Persian Gulf and eastwards to the Straits of Malacca. There was, in fact, an 'Empire of the Raj' until at least the First World War, in which Indian foreign policy interests were powerfully expressed and represented in the Gulf and on the Arabian and Swahili coasts, often in conflict with other British imperial interests.

The building blocks of British supremacy in the Indian Ocean colonies and bases, trade – treaties, humanitarianism, communications, naval warfare and imperial defence strategy – are discussed. The truism that the Indian Ocean was a 'British lake' is examined, as is the importance of Indian Ocean regional power centres, like India, South Africa, and Ceylon, that were part of the structure of imperial rule and mobilization in war and in peace. The second half of the chapter concentrates on the pre-1941 balance of power in the Indian Ocean, and the strategically vital though often neglected role of the Indian Ocean during the Second World War. To do this, naval history is blended with standard imperial history to present a clearer picture of activities in the region, and of their significance.

Though some notable works have concentrated on the Indian Ocean as a distinct region in the study of British imperial history (Gerald Graham's *Great Britain in the Indian Ocean* stands on its own) it can justly be labelled a region neglected by imperial – as opposed to regional – historians. The recent multi-volume *Oxford History of the British Empire* pays little attention to the Indian Ocean and sticks to the conventional land-region focus that has characterized the study of the empire. The Indian Ocean usually appears as a thoroughfare rather than a region in its own right, with links peculiar to itself, and in the history of the global war of 1939-1945, it is a poor relation to Europe, the Pacific, and the Middle East - many indeed would not even realize that the Indian Ocean was a front line theatre of war.

Components of the British Indian Ocean World

The Indian Ocean was a region of huge significance for the British imperial world economy and its strategic integrity for two centuries. It encompassed key sea routes and ocean annexes like the Persian Gulf, the Red Sea, and the Bay of Bengal, and kept the eastern empire in touch with the European heartland. It united Africa, Asia, Australia, and the Middle East and contained within these land borders was a galaxy of islands and atolls. Alongside relatively well-known islands like Ceylon, Mauritius, and Madagascar, was the Mauritian sub-empire of Rodrigues, the Chagos Archipelago, the Agalega islands, and the Cargados Carajos group. Until 1903, the Seychelles, itself a sprawl of many islands, was also ruled from Mauritius. Towards India and the east lay the Laccadives and the Maldives, the Andaman and Nicobar Islands, and the Cocos and Keeling Islands. Off the East African coast was Zanzibar and the Comoro Islands, off the Arabian coast were the islands of Socotra and Masirah, and the seldom-visited French islands of Kerguelen and Crozet in the distant Southern Ocean. In the colonial period, all had their strategic value, and all, if not British, presented an opportunity for a rival power to challenge Britain's commercial and naval preponderance. The Ocean's rim was also controlled in the main by Britain, with their colonial and semi-colonial holdings in Africa, Arabia, the Persian Gulf, India, Burma, South East Asia, Australia, and even in the Antarctic. (The British had also held sway at one time or another in the Dutch East Indies and French Indo-China).

What were the makings of the British Indian Ocean world, or the Anglo-Indian Indian Ocean world as it might more accurately be dubbed? Sea routes and commerce with the east were the bedrock, accounting for much of the inter-European warfare and land acquisition in the region.

India was a source of conflict between Britain and France, on land and at sea (for Robinson and Gallagher the partition of Africa was 'a vast footnote to the Indian empire', rooted in Indian defence concerns). The *parvenu* British presence in the region overlay centuries-old connections across the Ocean, as creeping imperialism transformed an initial contact of equals into eventual British rule or overlordship, for example in the domains of the Sultan of Zanzibar and Muscat, and in the Moghul province of Bengal. A growing imperial trading system was advanced in the nineteenth century by communications and transport developments, and the suppression of slavery and piracy drew Britain into further Indian Ocean entanglements.

Many different types of imperial expansion were in evidence in the Indian Ocean. It was an arena for private interest trade and sub-imperialism (the East India Company and the Imperial British East Africa Company), man-on-the-spot forward imperialism (Sir Stamford Raffles), business imperialism (Sir William Mackinnon), strategic imperialism (Suez and the Red Sea), military imperialism against European rivals (the Cape, Mauritius, and Ceylon), military imperialism against indigenous kingdoms (Kandy), and the pursuit of the civilizing mission (anti-slavery pressure on the region's leaders). Forged in an age of sail and spice, successive technological developments bound Indian Ocean ties tighter and enhanced the Ocean's strategic significance, like submarine telegraphy (the cable from the Cape to Australia touched on Indian Ocean islands), steamships (facilitating, for example, the Bombay-Suez monthly mail with a stop on the Arabian coast), wireless telegraphy (Cable & Wireless outposts were opened on numerous Indian Ocean islands), oil-fuelled engines (supercharging the importance of the Gulf), aircraft carriers, and submarines. From the Seven Years' War to the Cold War the Indian Ocean and its sea routes were of major strategic importance to Britain.

Strategy, Naval Warfare, and the Suppression of Piracy and Slavery

War between Britain and other European powers was a major impetus for an imperial presence, and for imperial conquest, in the Indian Ocean. The British had maintained a fleet in the East Indies since 1744 in support of the East India Company's trading settlements in India and the Malacca Straits. Long after French power in India had been eclipsed, the threat posed by French naval vessels and privateers remained serious and jeopardized British command of the sea. Until Ceylon and the Cape were taken from the Dutch, and Mauritius – a stronghold for pirates and French fleets sent from metropolitan France – was conquered by Indian arms in

1810, the security of British trade depended upon merchantmen forming convoys protected by Indian and British warships. (The convoy system was resuscitated during twentieth century conflicts.) The islands and atolls of the Indian Ocean together with the coastlines that frame it, provided bases during successive British struggles to maintain supremacy at sea and keep vital trade routes open.

Victory over European rivals in the wars of the eighteenth and nineteenth centuries left Britain master of the Indian Ocean and the foremost European power in India and the Far East. Through war prizes and new annexations the British gathered to themselves the great maritime gateways guarding the Indian Ocean – the Malacca Straits, the Suez Canal, and the Cape, along with almost all that lay between – bases that were either needed for refuelling and refitting British ships, or that simply could not be left in enemy hands. Some annexations were deliberate, others, like the Seychelles in 1794, were a by-product of campaigns elsewhere. Informal means were also employed to further imperial interests in the region and thwart those of Great Power rivals. For example, supporting and courting (increasingly through financial investments) the regime in Tehran was standard policy until 1914, as British and Indian officials regarded Persia as the key to stemming Russian southward expansion, defending India, and maintaining British paramountcy in the Persian Gulf. Local factors could also draw in the military arm of the British empire. The Indian Ocean became a major theatre of Royal Navy operations aimed at the suppression of the slave trade between East Africa and Arabia, and the activities of pirates in the same region required additional naval forces in order to protect merchant vessels and insure the safety of the sea routes upon which Britain's maritime trading empire had come to depend.

In this maritime trading empire, ports and coastal footholds were as connected to the ocean as they were to the land. Indeed, in the Indian Ocean, key imperial possessions were gained for purely oceanic reasons, with little concern for the hinterland entanglements that time would bring – one thinks of Aden, the Cape, and Ceylon. They were the nuts-and-bolts of an Empire in which 'the Royal Navy was neighbour to every country with a coastline', as were the ships of the British merchant marine, the Bombay marine, and (until 1863) India's own navy. The growing British naval presence led to an unparalleled knowledge of the region. By the mid-nineteenth century 'there were few coasts between the Cape and Malaya that had not been subject to exacting scrutiny' by the British. By then, the Persian Gulf was a familiar haunt, and expeditions like that sent by the Admiralty to survey the East African coast in 1822 were building an unrivalled oceanic knowledge. By the end of the century, the Indian

Ocean 'had become the most Europeanized of all the world's seas, not excluding the Mediterranean'.

Aside from Great Power rivalries, Indian Ocean expansion was undertaken to counter local powers whose ambitions threatened British interests (particularly if supported by a Great Power). Egypt's thrust towards the Persian Gulf threatened the overland route to India, prompting British action. On the other side of Arabia, the Egyptian threat, encouraged by France, led Britain (or rather, Indian troops from Aden – Aden was an outpost of the Bombay Presidency, then the Government of India, until 1937) to occupy Perim in 1857, a cork rammed into the neck of the Red Sea bottle. Similarly, Austrian and Italian designs on the island of Socotra were thwarted, and Ottoman expansion down the Arabian peninsula caused a British preserve to be more forcefully staked out in the hinterland of Aden. Egyptian expansion led to Aden-based Indian troops landing in Somaliland, an area considered important because of its role in provisioning Aden. Towards the end of the century the challenge of growing German influence in Zanzibar and East Africa prompted extensive British activity on the Swahili coast and its hinterland, and, in 1883, London took control of the Indian consulate on Zanzibar.

Trade, Shipping, and Cables

From the early modern period the Indian Ocean witnessed the growth of European trading companies as part of an emerging economic system marketing the commodities of Asia throughout the world. The East India Company created a factory system and administration stretching from the Red Sea to the Persian Gulf, South Asia, the Indonesian archipelago, and the China seas. Trans-oceanic trade was the central plank in British interests and operations in the Indian Ocean. Umbilically connected, trade brought an absorbing interest in shipping routes and their defence, and imperial defence concerns brought coaling stations, ports, and garrisons. These in turn became focal points of power, themselves emitting expansionist sub-waves.

Global trade was a hallmark of the British empire, and the Indian Ocean region was a key transport terminus and production centre fuelling its growth. Ceylonese coffee, Mauritian sugar, Burmese rice, Bengal jute, East African cloves and spices, Indian cotton textiles, Assam tea, and Chinese silk all became standard exports. Not only was trade *across* the Indian Ocean important to the British empire, but also trade *within* it. A burgeoning regional trade long pre-dated the British arrival. Zanzibar was an entrepôt for trade routes stretching far into Central Africa, and the western corner of a venerable triangular trade connecting Africa with the Persian Gulf and Bombay. Many Indian Ocean territories relied upon imports from elsewhere in the region. For example, Mauritius, like many other colonies, was a wholly agricultural country, yet one incapable of feeding itself. It exchanged its sugar for food, manufactures and consumer goods, and was therefore dependent upon seaborne imports, and, in times of war, faced starvation. This was the case when it was blockaded by the British during the Revolutionary and Napoleonic Wars, and during the Second World War when shipping shortages and enemy activity threatened to destroy Britain's capacity to supply the islands.

The Indian Ocean was lashed firmly to the Western world by the communications revolution of the nineteenth century. The laying of submarine cables linked Europe to India and Australasia by the last quarter of the nineteenth century. New shipping routes were developed and more Indian Ocean ports received scheduled steamer stops from the ships of companies like the Peninsular & Oriental Steam Navigation Company (P&O). Sir William Mackinnon, one of the great shipping magnates of the nineteenth century, built up a dominant position in the steamship world of the Indian Ocean through his British India Steam Navigation Company, winning government subsidies to carry mail. By subsidizing shipping lines, the British government gained the right to requisition ships in emergencies, particularly for the transit of troops. In 1869, the Indian Ocean was touched by one of the most remarkable feats of nineteenth century engineering, the cutting of the canal through the isthmus of Suez that united the oceans of Europe with those of Asia. This strengthened the links between Europe and Asia and supplemented the overland route to India via Egypt and the Red Sea that had blossomed since the 1830s.

Migration

The movement of people had for centuries bound the Indian Ocean region together, and migration continued apace in the age of British predominance as the burgeoning global economy made huge labour demands throughout the region. The slave trade flourished between the Swahili coast and Arabia, and Portuguese East Africa and Brazil. The demands of the sugar industry took many Africans from the east coast and Madagascar to Mauritius and Reunion. After the abolition of slavery in the British Empire came the massive flow of indentured labour from South India to Mauritius, and Indians also settled in large numbers in British East Africa, South Africa, and Malaya. Until 1866 Mauritius received more of India's indentured labourers than anywhere else in the world. Over 30,000 Indians built and worked the railway in Uganda (2,500 dying in the process), and many Chinese were recruited to work in the gold mines of the Transvaal. Between 1834 and 1920, 35 per cent of indentured labourers went to Mauritius (452,602), nine per cent to Malaya, and seventeen per cent to East Africa, Natal, and the Transvaal. Providing more specialized services, Sikhs and Muslims from the Punjab were recruited to serve as policemen in Hong Kong, Perak, the Straits Settlements, Nyasaland, Kenya, Uganda, and Somaliland. The Indian Ocean also experienced white settler imperialism, most notably in Mauritius.

Aden had for long been part of India's western trading empire with strong links to Gujarat. The East India Company had first surveyed Aden in 1609, but chose to develop its Red Sea activities at Mocha. Aden was taken by Indian forces in 1839, and became very much an Indian province. Indians were so numerous in East Africa that calls were heard from the Government of India and the Indian political class to make it an Indian territory after the First World War.

Regional Power Centres

Though the Indian Ocean was part of an imperial system of power and an emerging global economy, it was not just as an imperial thoroughfare and appendage. Thomas Metcalfe uses the term 'periphery-as-metropole' to emphasize the need to focus on the Indian Ocean in its own right. Such a perspective encourages us to look from the Indian Ocean 'out', focusing, in this instance, on imperial bases and their significance within the region and beyond. London was never capable of organizing the Empire alone, whether in peace or war, and depended absolutely on the skill and resources of regional imperial strongholds (what here are termed 'regional power centres'). Colonies, cities, and military commands helped the imperial centre administer, defend, and supply the empire, and co-ordinate its operation in times of war. Individual colonies and imperial cities like Bombay, Delhi, Cairo, Colombo, and Mombasa became regional power centres commanding their own resources and pursuing imperial interests in their respective regions. Imperial organizations like Middle East Command and the Middle East Supply Centre had responsibilities and powers ranging across vast regions.

One manifestation of the regional power centre was the operation of sub-imperial forces in furthering British imperial interests and responsibilities, often without metropolitan sanction (for example the Governor of Mauritius' ambitions in Madagascar). India was the greatest regional power centre, second only to Britain itself as the source of imperial power. India had always been a strikingly autonomous branch of the imperial enterprise, emitting expansionist impulses at its own behest. The East India Company was a private government, its Governor-General a despot marshalling a large army and navy and formulating expansionist policies independent of London. Even its regional governors, like Sir Bartle Frere in Bombay, had widespread regional responsibilities, in his case stretching to the Persian Gulf, the Red Sea, and the Swahili coast. Expansion into the Persian Gulf was driven from India, to protect trade against piracy, to prevent Arabs ceding territory to rival Great Powers, to protect shipping, and to preserve the internal peace and safety of land communications with India. British India had an empire of its own, including Burma, Aden (both until the 1930s), the Settlements of Penang and Malacca (transferred to the Colonial Office in 1867), as well, briefly, as Ceylon.

For three centuries, India projected British power and interest in the region. It was the East India Company's ships, *Ascension* and *Cape of Good Hope*, that first made landfall on the Seychelles on behalf of Britain in 1609. Ten thousand Indian Army troops conquered Rodrigues and Mauritius in 1809-10, 8,000 Indian troops passed through the Suez Canal destined for Malta in 1878 at a time of Anglo-Russian tension, 15,000 Indian troops were sent to discipline the Emperor of Abyssinia in 1869, and Indian troops formed the bulk of the columns that marched across the North West Frontier into Afghanistan in 1839 and again forty years later. Indian Army officers were pioneers of African exploration, and Indian ships plied the maritime trade routes and looked to their security. In a later period, Indian troops made up the bulk of Second World War armies like the Fourteenth Army in Burma, the Tenth Army of Persia and Iraq Command, and Indian divisions served in the Eighth Army in the Western Desert.

A British Lake?

'Almost a British lake', wrote one imperial defence expert more accurately if more pedantically, because Britain did not own, occupy, or influence every bit of dry land either framing the ocean or forming the islands within it. However, the 'almost' could not hide the fact of overwhelming superiority in the region in the halcyon days of British naval hegemony, before aircraft carriers and the proliferation of naval rivals blew apart the old benchmarks of power founded upon capital ships and bases. Britain garrisoned or influenced most of the lands around and within the Indian Ocean, and none were immune from its naval power. Its territories were largely secure from land-based attack, no other power had a first class naval base in its waters, the cable system was British-controlled, the carrying trade was mainly British, and Britain held the great 'maritime gateways' through which access would be needed by any power planning to challenge Britain's position. The French, of course, held Madagascar with its Grand Harbour at Diego Suarez, but it only served to illustrate the problem of having a first class naval base without a significant naval force to occupy it. In times of war, great bases count for little without command of the sea, as the British discovered at Singapore in 1942, and as the French discovered when Madagascar was taken by the British in the same year. In persuading the French Chamber of Deputies to vote the estimate for the construction of the Diego Suarez base, the Minister of Marine said that it would hand to France 'command of the Indian Ocean'. But it did not.

Another reason for urging caution in the application of the 'British lake' axiom is that though one might talk glibly about British hegemony in the Indian Ocean, it was only hegemony of a sort. It was a hegemony partly founded upon the *absence* of challengers rather than on overwhelming British strength. Before 1869, British dominance was essentially unchallenged, but the opening of the Suez Canal changed that. Even in the period before the rise of challengers to the Royal Navy's supremacy, though the Admiralty might maintain an establishment capable of preventing a global maritime challenge from another Great Power, the requisite strength for this was often ineffectual in pursuing other traditional naval concerns of a more localized nature, like the suppression of piracy and the slave trade. After the Revolutionary and Napoleonic Wars the size of the fleet plummeted from 230 ships-of-the-line in 1814 to forty nine in 1820 (excluding smaller vessels), and by the early 1840s the combined strength of the Cape and East Indies Squadrons was less than twenty ships. In those decades of European peace and unchallenged British ascendancy, this small number of ships was enough to keep watch on the French ports and the major sea routes of the world, but not to hunt for needles in needlestacks and make all slavers and pirates quake with fear. However, as Graham remarks, 'the presence of even two or three sloops cruising the Malacca Straits or along the west coast of the Malayan Peninsula bolstered confidence among the trading community . . .

and certainly discouraged any major operations by pirates . . . A small emergency naval force, the East India Squadron . . . was insufficient to quell the ubiquitous pirates and slavers, but as a symbol it was potent enough to keep the Indian Ocean a *mare clausum* to major hostile forces'.

Finally, one might take issue with Panikkar's assertion that because the Indian Ocean was so well established as a British lake, 'the question of sea power did not arise . . . and no one was interested in discovering the relations of the sea to Indian Defence. In the result, the entire emphasis was on the land frontier and Indian Defence was equated with the maintenance of a powerful army on the North-West Frontier'. Imperial strategists would have disagreed absolutely; imperial defence was founded upon the security of the seas and particularly the security of the sea routes leading to India, and its defence – until the Second World War altered perceptions dramatically – was, like Britain's, dependent upon capital ships and supremacy at sea. As Panikkar himself writes later in his book, 'control of the Indian Ocean was a corollary of the mastery of the Atlantic'. Until the rise of first class naval powers in Europe and the Pacific, 'the mere existence of the Grand Fleet was enough to secure the safety of India'. Furthermore, one should not forget that, in addition to protecting the North-West Frontier, the Indian Army was a projectile that sea power could, and did, hurl across the Indian Ocean and beyond in pursuit of British and British-Indian military imperatives.

The Twentieth Century to 1941

In the years of naval competition leading up to the First World War, the foundations were laid for the near-fatal challenge to British hegemony in the Indian Ocean that was to come in the 1940s. The German threat in European waters and the meteoric rise of Japan forced Britain to renounce paramountcy in the Far East, and enter into the Anglo-Japanese alliance of 1902. Thereafter, Britain's Far Eastern holdings were secured by the Royal Navy in alliance with the Japanese navy, as capital ships concentrated in Home Waters to face the threat of the German High Seas Fleet. Inevitably, this veiled retreat had direct and serious repercussions for the Indian Ocean. By 1914, 'with greater naval forces than she had ever before possessed [Britain] was compelled to renounce or farm out her overseas commitments . . . Atlantic supremacy was no longer identical with two-hemisphere supremacy'.

The relatively easy British hegemony established in the Indian Ocean after the Revolutionary and Napoleonic Wars was challenged during the First World War, even though the German navy did not pose a grave threat to British control, and the Royal Navy successfully blockaded German colonies in the region. Since annexations in the 1880s Germany had been an African and Indian Ocean colonial power. Although the main burden of Germany's naval threat was in Europe, victory there would have led to sweeping German victories in the other oceans of the world where the white ensign held sway. German commerce-raiding operations in the Indian Ocean were intended to occupy British forces more urgently needed elsewhere by threatening British possessions and the security of shipping lanes. After entering the Indian Ocean the *Emden*, a cruiser from the German Pacific Squadron, sank or captured sixteen ships, raided Penang in Malaya, bombarded oil tanks at Madras, and kept a British battleship and six cruisers busy in pursuit. She was eventually sunk by HMAS *Sydney* in November 1914 off the Cocos-Keeling Islands where she had shelled, and deposited a raiding party to attack, the Cable & Wireless station. Similarly the *Konigsberg* attacked British shipping at

Zanzibar, where she sank HMS *Pegasus,* and led British forces a merry song-and-dance until eventually hunted down in East Africa's Rufiji Delta by shallow-draft monitors sent out from Britain for the purpose.

The Indian Ocean region also witnessed a good deal of land warfare during the First World War. The conflict between British imperial forces and those of the Ottoman Empire saw British advances along the Tigris and Euphrates rivers in Iraq (featuring in 1916 the surrender of a 9,000-strong British-Indian force at Kut). The protracted war in East Africa came to pin down over 100,000 imperial troops from Britain, India, South Africa, and West Africa, and hundreds of thousands of military labourers from East and Central Africa. At the outbreak of war, German wireless stations, like that at Dar-es-Salaam, threatened British shipping by providing intelligence to German warships. Imperial politicians like Leo Amery and Lord Alfred Milner did not allow events in Europe to cloud the broader global picture, and were determined to take the German possessions in the southern hemisphere.

The First World War demonstrated Britain's capacity to mobilize the resources of its Empire and move them around, a telling benefit of sea power. Even the smallest colonies pitched in. Mauritius, for example, provided a Labour Battalion of 1,700 men that was shipped overseas to work in Mesopotamia in the campaign against the Ottoman empire, and over 500 white Mauritians volunteered for the British and French armies on the Western Front. The island became part of the wireless telegraphy chain across the Indian Ocean, and raised contributions for the war effort by voluntary donations, a special tax on sugar exports, and gifts from government surpluses. Among other things, the island purchased thirty aircraft for the British forces.

Victory in 1918 removed the German threat to the Indian Ocean, and Britain gained new imperial holdings in the region, like Iraq and Tanganyika. Oil was a growing British interest reaffirming the importance of British sway in the Indian Ocean and Persian Gulf. British power in the Indian Ocean remained unshaken, and it is little surprise that in the inter-war years traditional formulations of imperial military geography and imperial defence prevailed. Indian Ocean security continued to be predicated on the assumption of British command of the sea, ultimately based on capital ships, in tandem with penny-packet garrisons. The threat of carrier-borne aircraft to these traditional imperial defence assumptions was not appreciated until too late. But the rise of Japanese and American naval power had fundamentally altered the situation anyway, along with the sheer proliferation of potential enemies wielding navies large or small. *Pax Britannica* had been based upon a virtual monopoly of naval might, and though the Royal Navy might still be the largest navy, it was certainly no longer the only one, and so naval hegemony – and with it the fundamentals of the classic formulation of imperial defence based on sea power – had vanished. Elsewhere in the Indian Ocean, the inter-war years saw the French pour money into the naval base at Diego Suarez, and Mussolini attempt to make Massawa in Eritrea a naval base capable of having a decisive impact in the region of the Suez Canal artery.

Regional Power Centres and the Second World War

Between 1939 and 1945 Britain, unlike Germany, Japan, and Italy, simultaneously fought a European and an Asian war. As a result of its imperial history Britain fought a very imperial war in terms of battle zones (for example the Middle East, Burma, and the Indian Ocean), and in so doing relied on the resources of many colonies – as bases for fleets and armies, as sources of military and civilian labour, and as sources of food and raw materials vital for the world wide war effort. Britain's was a unique war, fought alongside imperial allies mainly in centres of traditional imperial military and naval power. This global war effort depended on the large-scale oceanic transfer of resources from one theatre of operations to another, and this relied upon command of the seas. Given the global nature of Britain's war effort, there was a need for regional power centres to command and co-ordinate military, civilian, and raw material resources and to serve and protect lengthy supply routes. Far from London being the only brain in the monumental operation of bringing the British imperial world to war, there were many regional nerve centres upon which it depended.

The 'periphery-as-metropole' (or 'regional power centre') lens referred to earlier encourages examination of the regional military and naval command structures used by the British in the Second World War. An example is provided by East Africa Command, established in 1941 to relieve the massive Middle East Command of an area of responsibility. This was an army command responsible for mobilizing regional resources from all British colonies in East and Central Africa and for prosecuting the war against the Italians in Abyssinia, Kenya, Somaliland, and the Sudan. It was later charged with the task of invading Madagascar, thus preventing the Japanese from entering the western Indian Ocean, and was responsible for garrisoning Mauritius, Madagascar, and the Seychelles, providing an Army contribution to what was primarily a naval defence of British territories in the Indian Ocean.

Regional power centres were important in running the Empire, especially in war-time. Examples of colonies acting as regional power centres in the Indian Ocean during the Second World War are provided by South Africa and Ceylon. By the Second World War, South Africa was a regional power of growing significance, economically invaluable to the empire and occupying a key strategic position. South African army and air forces played a leading role in British military operations in East Africa, North Africa, and the Indian Ocean in two world wars, and Jan Smuts was a long-standing confidant of Churchill and a powerful voice when it came to the direction of the war in the region, for example urging the occupation of Madagascar, duly accomplished by British, East African, and South African troops. The South African economy produced munitions, gold, minerals, and war materials in great quantity as it responded to the demands of war and became increasingly industrialized. The threat of Japanese invasion was taken seriously, barbed-wire and crocodiles' teeth defences appeared on South African beaches, and the government planned to arm Africans should landings take place. The war stimulated South Africa's interest in what was termed 'seaward defence'. Minesweepers, trawlers, and whalers were taken into service to create a South African navy capable of coastal defence and patrol tasks in a region of great importance for Allied shipping, and even of sending vessels to aid the Mediterranean Fleet. Forced to look increasingly to its own resources as the imperial defence umbrella showed signs of weakening, 'some South Africans began to realize that the Union was an Indian Ocean power'.

On the other side of the Indian Ocean, Ceylon was equally important as an imperial base and strategic outpost. It was a major base for the Royal Navy from the late eighteenth century, and during the Second World War was home to the Eastern Fleet, from late 1941 until late 1944 Britain's sole naval weapon east of Suez. After the loss of Singapore, Ceylon found itself on the front line and became the most important naval base between the Cape and Australia and a key to the defence of the Indian Ocean. The Eastern Fleet was responsible for keeping the Indian Ocean a 'British Lake' – particularly by deterring Japan from advancing westwards in force – and for defending the convoys bearing war materials and men across the ocean. Ceylon was an assembly point for ships travelling in convoy. It was also a centre for British air power covering the Indian Ocean and South Asia, and a barracks for British, Indian, African, and Australian troops breaking trans-oceanic journeys or destined for the Burma front. With the arrival in Ceylon in 1944 of Admiral Lord Louis Mountbatten's South East Asia Command (SEAC), the island became a concentration point for the re-invasion of Japanese-occupied territory in the East Indies and South East Asia.

On the naval front, Ceylon recruited its own Royal Naval Volunteer Reserve that assumed responsibility for convoy duties between ports in Ceylon and India, and a mine-sweeping role in local waters. It was also the site of a major refitting and repair facility for naval and merchant vessels. The British intelligence-gathering network in Asia and the Far East, centred on Bletchley Park, maintained an important station in

Ceylon known as Far East Combined Bureau, and Special Operations Executive (SOE) used Ceylon as a base for operations in neighbouring Japanese-occupied territory. Finally, the island provided indigenous soldiers for home and regional defence, raising 26,000 men for service in military units, and more for an Essential Services Labour Corps to ensure a smooth supply of labour for the extensive military construction that took place on the island, particularly after the arrival of SEAC. Its arrival led to a massive demand for local civilian labour, particularly for the construction of airfields. Superbly illustrating the impact that participation in a world war could have on a domestic economy and society, these labour demands eradicated the unemployment problems that had characterized the inter-war years, by employing over 83,000 people.

Even the smallest colonial territories provided military manpower for the Empire's war effort. Other examples from the Indian Ocean region include Aden and the Seychelles. On the outbreak of war the Aden Protectorate Levies were placed under RAF command and used to guard airfields in the colony, and to garrison Socotra and Masirah islands. Their total strength in May 1945 was 1,800. In addition, an Aden Pioneer Corps was raised and placed under Air Ministry control. A small Defence Force was raised in the Seychelles after the outbreak of war and placed under War Office control, and two companies of Pioneers (totalling 854 men) were sent to the Middle East. All together over 1,500 Seychellois served in the forces.

The Indian Ocean, the Eastern Fleet, and the Royal Air Force

The Indian Ocean was of immense importance to Britain and its allies during the Second World War for the same old reasons. Its key islands and the land massed around its rim had to remain under British sway in order to safeguard vital convoy routes. Even its tiny, barely inhabited islands were to feature in the conflict, as they were used by enemy vessels as hideouts and supply bases, necessitating lonely patrols by British warships and aircraft trying to locate them. The Eastern Fleet relied on the region's harbours, like Trincomalee, Colombo, Mombasa, Diego Suarez in Madagascar (captured from the French), Port Louis in Mauritius, and its secret fleet base at Addu Atoll in the Maldives. The Seychelles became a focal point for monitoring shipping 200 miles to the north on a passage between the Far East, Ceylon, East Africa, and the Cape, and six to seven hundred miles to the west traffic converged between the Cape and the Red Sea, and ships returned to the Far East (the Seychelles were also a connecting point on the cable and telegraph lines between Zanzibar, Aden, Mauritius, and Ceylon).

Indian Ocean convoy routes transported hundreds of thousands of troops and all manner of essential supplies from Britain to the Middle East and South Asia, and from the Middle East to Australasia, as well as bringing similar cargoes in the opposite direction and supplying the food requirements of import-dependent colonies like Mauritius and Ceylon. With the virtual closure of the Mediterranean-Suez route (and the cutting of the Mediterranean cables) the importance of the Cape route increased for the supply of the major British war effort in the Middle East and Mediterranean. Another vital Indian Ocean convoy route took Lend-Lease supplies to Russia via the Mozambique Channel and the Persian Gulf, from where British oil supplies entered the Indian Ocean.

Throughout the Second World War the Indian Ocean was a theatre of sustained activity for Japanese and German (and to a lesser extent, Italian) submarines and surface raiders. Despite the efforts of the Royal Navy and the Royal Air Force (RAF) Indian Ocean sinkings throughout the war were high, amounting to 385 British, Allied, and neutral ships, totalling 1,787,870 tons. Early in the war the main threat came from disguised German surface raiders, like the notorious ships *Pinguin* and *Atlantis*, both sunk in 1941 by British cruisers (the pocket battleship *Graf Spee* also appeared in the Indian Ocean before its destruction off Montevideo.) Then came the U-boat period, accompanied by their much-hunted supply vessels, like the *Charlotte Schleimann* and *Brake*, both sunk south of Mauritius through the combined efforts of Eastern Fleet destroyers and cruisers, RAF flying boats, and ULTRA intelligence. Later Japanese I-boats also operated extensively in the Indian Ocean, surveying all British coastlines in the region. Most of the ships sunk by the German and Japanese submarine packs that operated in the Indian Ocean were merchant vessels. As in other theatres, the British were particularly successful in protecting the troopships that shunted hundreds of thousands of men around the world. However, one of the handful of disasters that befell British troopships occurred in the Indian Ocean in February 1944, when the SS *Khedive Ismail*, part of the five-ship convoy KR8 bound from Mombasa to Colombo, was sunk by a Japanese submarine with the loss of 1,134 lives. Most of the dead were men of the 301st Field Regiment, East African Artillery, destined for the war in Burma.

Before the entry of Japan into the war in December 1941, the British position in the Indian Ocean was worrying but not critical. However, after Pearl Harbor, the main strategic challenge to Britain's position in the Indian Ocean was posed by the Imperial Japanese Navy. Following the totemic sinkings of HMS *Prince of Wales* and *Repulse* off Malaya in December 1941 and the Allied defeat at the Battle of the Java Sea, the Eastern Fleet based at Ceylon was Britain's only naval presence east of Suez, and was faced by what temporarily was 'the most powerful navy in the world'. At this gravest point in the eastern war the Eastern Fleet came into existence, an umbrella for the remnants of all British naval forces east of Suez (the name East Indies Fleet was restored in November 1944 when the Eastern Fleet was split, the most powerful elements going to form the new British Pacific Fleet that provided the British with a potent naval re-entry into the Pacific.)

The Eastern Fleet was expected to patrol and defend a vast area; the Commander-in-Chief at naval headquarters Colombo (Admiral Sir James Somerville) had subordinate officers at naval shore bases in India, East Africa, the Persian Gulf, Aden, Mauritius, the Seychelles, Bombay, Calcutta, Cochin, Addu Atoll, Hormuz, Shatt-el-Arab, Bahrein, and Kuwait. The Eastern Fleet's job was to protect convoys crossing the Indian Ocean, to hunt and kill enemy submarines and raiders, and – after Pearl Harbor and the raid on Ceylon – to evade being destroyed by the Imperial Japanese Navy whilst trying to deny the enemy the free use of this vast ocean. The Eastern Fleet was also responsible for blockading enemy territories – like the Indian Ocean islands of Vichy France – for intercepting Vichy convoys and the submarine traffic in technology and industrial raw materials between Japan and Germany, and for dealing with the Italian threat in the Red Sea, centred on the port of Massawa. The Fleet provided air cover for Army operations, like that in Iraq in 1941 that overthrew Rashid Ali, and that dispatched for the capture of Vichy Madagascar in 1942 (for this campaign the Eastern Fleet was reinforced by units from Force 'H', the powerful squadron based on Gibraltar).

At the height of the Japanese threat in 1941-42 the Eastern Fleet had to accomplish its manifold tasks with limited resources. For example, four of its five battleships were of First World War vintage, slow, and lacking armoured decks – in other words, easy pickings for Japanese carrier-borne aircraft. It lacked fleet training, sufficient firepower, modern aircraft, and adequate harbour protection at Trincomalee and Colombo. Though the Eastern Fleet was large on paper, it was simply no match for the five carriers and four modern battleships that formed the core of Admiral Nagumo's fleet. In April 1942 Nagumo's carrier force raided Ceylon, attempting to Pearl-Harbor the Eastern Fleet and secure Japan's western perimeters after its four months of conquest and expansion. Though he failed to destroy the main units of the Fleet, the raids cost over 1,000 British and Ceylonese lives, two cruisers, a carrier, two destroyers, twenty three merchantmen, a number of smaller vessels, and over thirty combat aircraft. Following the raids Admiral Somerville withdrew the older units of the Fleet to Mombasa in East Africa – and 'in effect abandoned the east and central Indian Ocean to concentrate on the defence of the western seaboard with the vital supply route to the Middle East'. In the period before the waning of Japan's naval star the Eastern Fleet's main purpose was to *deter* the Japanese from striking across the Indian Ocean. It was a game of bluff in which, as Somerville put it, he had 'to lie low in one sense but be pretty active in another – keep the old tarts [the 'R' class battleships] out of the picture and roar about with the others'.

When British fortunes improved, the Eastern Fleet was able to return to its Ceylon headquarters in September 1943. However, 1943, though less dangerous than 1942, was a frustrating year for the Eastern Fleet because it was not strong enough to take the offensive, and had to content itself with its unglamorous but essential policing role in the Indian Ocean. The war in the Atlantic and the Mediterranean and the watch on the Channel ports, meant that other theatres had priority for scant naval resources. The Eastern Fleet often found itself denied ships newly-launched or newly-released from other theatres, or stripped of major units required more urgently elsewhere, for example losing its main carriers to the battle to supply Malta in 1943.

However, Allied victories in other regions – at Midway, in the Mediterranean and the Atlantic, the sinking of the *Tirpitz* – eased the pressure on British naval resources, and the Eastern Fleet began to be strengthened, preparatory to a British re-entry into the Pacific and the planned re-invasion of all Allied occupied territory under SEAC. With the tide turning against the Japanese, the Eastern Fleet was able to take proper offensive fleet action against Japanese land forces and installations for the first time. It bombarded military positions on the Andaman and Nicobar Islands (taken by the Japanese in 1942 after being abandoned by the British) and oil installations in Sumatra, and aided the land forces fighting in Burma, for example covering operations during the third Arakan offensive. The Eastern Fleet cleared Japanese minefields and carried out major sweeps of shipping routes used by vessels supplying Japanese land forces. It also laid minefields to disrupt Japanese shipping, usually with submarines from the new submarine fleet based on Ceylon (submarines laid 490 mines). The Fleet also attacked, where they presented themselves, Japanese surface vessels, for example sinking the heavy cruiser *Haguro*, the last major Japanese warship sunk during the war. *Haguro* was attacked in the Malacca Straits by the Eastern Fleet's 26th Destroyer Flotilla, working on intelligence derived from broken Japanese naval codes intercepted by Far East Combined Bureau in Ceylon. But the Fleet's main task – and arguably Britain's main contribution to the war against Japan, notwithstanding the final and crushing victory of Slim's Fourteenth Army in Burma – was in keeping the Indian Ocean sea routes open for the absolutely crucial convoys.

It should not be forgotten that the war in the Indian Ocean featured both the Royal Navy *and* the RAF. In the face of the mounting Japanese threat in 1941, Ceylon's air defences had been steadily augmented. The

Fleet Air Arm had a few squadrons of Fulmars on the island, and Hurricanes, Blenheims, and Swordfish were flown in from the Middle East and India or flown off of the Eastern Fleet's carriers. Some RAF aircraft escaped to Ceylon after the defeats in Malaya and Sumatra. Though still heavily outnumbered by the 130 planes that attacked the island in early April 1942, the forces available at least put up some resistance and ensured that the Japanese did not have an unopposed run as they came in to bomb Colombo and Trincomalee. Whilst hardly a significant reverse, this was the first time that the Japanese had experienced anything approaching an opposed raid and suffered losses. Based throughout the Indian Ocean, Catalina and Sunderland flying boats played a role patrolling the region and greatly extending the search capacity of the Eastern Fleet, as did the land-based bombers concentrated mainly on the Arabian peninsula. Another important RAF role was the provision of meteorological intelligence and photographic reconnaissance.

Ceylon became the home of 222 Group, responsible for aircraft scattered throughout the Indian Ocean region from East Africa to Aden and east as far as the Japanese perimeter. Applying lessons learned in the Battle of the Atlantic, in early 1944 Indian Ocean General Reconnaissance Operations (IOGROPS) was formed, a unified control centre for RAF units in the Indian Ocean region. By 1944, there were about 160 aircraft stationed in and around the Indian Ocean, half of which were flying boats based mainly in Ceylon, India, and East Africa, the other half land-based aircraft mainly in Ceylon, India, Aden, and the Persian Gulf. There were also many forward air strips and flying boat anchorages in frequent use, for example in Mauritius, the Comoros, the Maldives, the Andamans, Rangoon, Socotra, Kenya, South Africa, the Seychelles, Madagascar, and Somaliland.

The Home Front

'Total war' impacted the home front in the Indian Ocean as everywhere else in the colonial world. The war's intrusion was manifest in terms of military and civilian labour recruitment, inflation and profiteering, new job opportunities for women, the growth of 'big' government, air raid precautions, home guard formations, rationing, and inward and outward migrations of all descriptions (for example, bringing Italian prisoners-of-war to Ceylon and Jewish detainees to Mauritius). The war required every effort to increase production on the colonial home front, particularly of precious raw materials like rubber. After the loss of Malaya, Ceylon produced sixty per cent of Allied supplies.

There is not the space here to do more than highlight one facet of the war's impact upon the home front, the crucial issue of food supplies, on which civilian health and morale depended. The war disrupted regional networks of trade, and civilian populations throughout the region suffered food shortages and dietary change. Mauritius and Ceylon were dependent on seaborne imports, and suffered during times of war, whether when blockaded by the British during the Revolutionary and Napoleonic Wars, or during the Second World War when shipping shortages, blockade of neighbouring Vichy territories, and enemy vessels threatened to destroy Britain's capacity to supply the islands. It was indeed, 'one vast interconnected world'.

Mauritius relied on food imports from Burma, the Agalega Islands, Rodrigues, and Madagascar. It depended for meat on the enormous bullock trade of Madagascar, and on the Mauritian dependency, the Agalega Islands, for edible oils. These sources of supply were severed by the war given Madagascar's Vichy status, and the absence of a ship to keep Mauritius in touch with its dependencies. Vital supplies of Burmese rice were lost, a severe problem as it was the staple food of the island's Indian majority. The loss of rice imports led to rationing and dietary change, requiring Nutrition Demonstration Units to tour the villages teaching Indians how to make bread. Compulsory food production schemes were initiated by colonial governments aimed at reducing import dependence, but these schemes rarely met with much success. As an alternative to lost supply sources Mauritius came to depend on infrequent though telling shipments of foodstuffs from other regional centres, like Australia, India, and South Africa.

Another example of food interdependence within the region linked Ceylon and Egypt. In 1943 the failure of rains in Yemen and the Aden Protectorate coincided with a shortage of cereals in the Red Sea area, and an urgent demand for rice in Ceylon. This crisis was overcome thanks largely to the Middle East Supply Centre, a British organization that co-ordinated supplies for twenty territories in the Middle East, North Africa, and Arabia. It ensured that for the remainder of the war the bulk of Egypt's surplus rice was exported to Ceylon.

Conclusion

By late 1941 the Imperial Japanese Navy had achieved a position that would have allowed it to dominate the Indian Ocean, and Churchill considered the threat to Ceylon and Britain's position in the Indian Ocean in those crucial months to be 'the most dangerous moment' of the war. However, Japanese operations in the Indian Ocean represented a strategic defeat. This was despite tactical successes like the sinking of HMS Prince of Wales and Repulse, the torpedoing of HMS Ramilles and the tanker British Loyalty at Diego Suarez by midget submarines, the destruction of major British warships in the raids on Ceylon, and Vice Admiral Ozawa's Malaya Force excursion into the Bay of Bengal which sank twenty three merchant vessels. For Japan failed to detect the bulk of the British Eastern Fleet, let alone destroy it, failed to prevent massive convoys crossing the ocean, and failed to take Madagascar. Indian Ocean operations were the main German interest in the Japanese war effort in 1942, and German pressure on Japan to act decisively in the region increased after defeat at El Alamein in October 1942. However, by then, the Japanese high water mark had been reached, and the Indian Ocean had become 'a sea too far'.

Though stretched, the Eastern Fleet and RAF had a strategic conception of the Indian Ocean as a whole. This stands in contrast to the Axis failure to push home advantages in the region that could have profoundly affected the war's outcome, and to coordinate their military efforts effectively. Germany and Japan recognized the great significance of the Indian Ocean, particularly the high commands of their respective navies, but were never able to overcome jealousies and translate their awareness into effective cooperative action to extinguish British forces in the region. The Indian Ocean was one of the great might-have-beens for the Axis.

The British had something of a tradition of imperial warfare. A local and global view of the world's oceans and land masses was evinced by British political and military leaders in numerous world wars, and by their subordinates. They were schooled to view the world in imperial terms and to understand how, for example, the Cape related to the defence of Australasia, Ceylon to the security of the Persian Gulf, and sea routes and naval bases to the very survival of the empire. Even when faced with the prospect of Japanese ascendancy in the Indian Ocean and possible defeat in the Middle East, the British succeeded in acting and thinking imperially, sending scarce resources, often in minute quantities, to protect tiny islands or to at least give the show of doing so; flying in penny-packets of Hurricanes and obsolete biplanes to defend Ceylon against the might of Nagumo's carriers; constructing dummy wooden guns and radar installations on the island to fool Japanese aerial reconnaissance; installing a brace of ex-naval 6-inch guns at Port Victoria in the Seychelles, manned by Ceylon Garrison Artillery troops; deploying an East African infantry battalion to beef up the defences of Mauritius; and sending Mauritian part-time soldiers to garrison Rodrigues. Luck, skill, and hard-won victories here and in other regions meant that, at the end of the war, the Indian Ocean was still a 'British Lake', patrolled by a fleet (the re-constituted East Indies Fleet) that comprised two battleships, 16 escort carriers, 13 cruisers, 43 destroyers, and over 130 other vessels. Britain remained the dominant power in the Indian Ocean, albeit in the new geostrategic setting of the superpowers and the Cold War, until the withdrawal from east of Suez in the late 1960s.

Select Reference List

Anderson, C. (2000), *Convicts in the Indian Ocean: Transportation from South Asia to Mauritius, 1815-53* (Houndmills: Macmillan Press).

Dockrill, S. (2003), *Britain's Retreat from East of Suez: The Choice between Europe and the World?* (Basingstoke: Palgrave Macmillan).

Graham, G. S. (1967), *Great Britain in the Indian Ocean: A Study of Maritime Enterprise 1810-1850* (Oxford: Clarendon Press).

Kearney, M. (2004), *The Indian Ocean in World History* (London: Routledge).

Louis, W. M. R., ed. (1998-9), *The Oxford History of the British Empire* (Oxford University Press).

Munro, J. F. (2003), *Maritime Enterprise and Empire: Sir William MacKinnon and His Business Network, 1823-1893* (Woodbridge, Suffolk: Boydell Press).

Pearson, M. (2003), *The Indian Ocean* (London: Routledge, 2003).

Scarr, D. (1998), *Slaving and Slavery in the Indian Ocean* (New York: St Martin's Press).

CHAPTER 3

The West in the Geopolitics of the Indian Ocean and India

Jean Houbert

Introduction

It is fashionable at the moment to say that, with the end of the Cold War, the transnational capitalist economy is globalizing and that the state, the interstate system and geopolitics will soon be redundant. 'Globalization', whatever else it may mean, refers to the globe, planet earth, which has been encompassed since the Age of Discoveries. This chapter argues that through the dynamic of colonization-decolonization the interstate system has expanded from its original base in Europe to become world-wide and that this global system of states remains as important today as the transnational capitalist economy. It has been argued that in the age of air power and nuclear weapons the strategic significance of the configuration of land and sea on the planet, that classical geopolitics had highlighted, is no longer pertinent. This chapter wants to show that with regard to the zone of the Indian Ocean in particular the configuration of land and sea remains highly significant in geopolitics. Projection of power from the sea to the land is taken to be sea power, whether it is the landing of an army, missiles fired from surface ships or submarines, planes from aircraft carriers or from offshore islands. On a larger, global, scale, projection of power from the sea can be taken to mean the projection of the power of the United States on to the rim of Eurasia to 'contain ' the Soviet Union during the Cold War and now to intervene, notably, in the 'Arc of Crisis ' on the northern rim of the Indian Ocean.

The Dynamic of Colonization-decolonization in Globalization

The Discovery of the Sea

The Indian Ocean as a body of seawater is a part of the global ocean. Right at the start of modernity, navigators from Western Europe searching for a sea route to India, to the fabulous riches of the Indies, discovered that the Indian Ocean was not a gigantic lake but was connected to the Atlantic. Indeed, the European sailors discovered that all the seas of the planet were interlinked forming a global watery highway on which it was possible to travel to all parts of the world. This took place very rapidly: when Columbus embarked on his historic voyage in 1492, Europeans had crossed no oceans at all; yet, only thirty years later, in 1522, one of the ships of Magellan's fleet returned to Europe having gone right round the world and crossed all the oceans. The Europeans' discovery of the sea was doubtless the central breakthrough of the Age of Discoveries that gave Europe the edge over the other continents. For no sooner had the Europeans discovered the sea then they imposed their hegemony over the global ocean. Innovations in maritime technology, most notably the building of artillery into the strong, nailed (instead of sewed), hull of ships, thus transforming them into floating fortresses, rather than being just platforms for the infantry, gave the Europeans a compelling advantage over the maritime activities of other civilizations. The gunned-sailing ships of Atlantic-Europe drove the ships of non-Europeans out of the seas of the world. As the sea makes up over two-thirds of the surface of the planet, the Europeans, through the exercise of sea power, held the world in their sway. This opened what Panikkar called the Vasco da Gama epoch of Indian history. The sea power, which the West acquired in these early days, has been retained in the present international system, with one of Europe's erstwhile settler-colonies, the United States, now playing the leading role. European hegemony over the sea was translated immediately into power over the land in the Americas. In Asia, when the Europeans first arrived on the coast, the indigenous civilizations were much too powerful on land to be conquered by a handful of conquistadors, thousands of miles from Europe. At first, the Europeans seldom ventured inland in Asia beyond the reach of their naval guns. Moreover, unlike what happened in America, in Asia it was the Europeans who had no resistance to diseases. Besides, the objective of the Europeans in Asia, initially, was trade rather than conquest. Power at sea, plus a few fortified trading enclaves on the coasts, was sufficient. However, rivalry among the European trading companies for monopolies in Asiatic goods, wars among European states, in which overseas trade were often one of the stakes, the erosion and eventual collapse of indigenous authorities, all drew the Europeans into political involvement in Asia. The time came when, with industrialiszation and modern medicine, the military power of Europe on land as well as at sea became irresistible. The only choice left to the Asiatic civilizations was to adopt the institutions – and, in particular, the modern state and the capitalist mode of production – which had made Europe rich and powerful, or have modernity imposed through European colonial rule. Japan was the only Asiatic state which succeeded in modernizing itself rapidly enough to join the ranks of the European colonial powers and, indeed, eventually to defeat them in Asia.

Colonization-Decolonization

To understand the part played by the dynamic of European colonization-decolonization in the globalization of the international political system, it is necessary to distinguish between two kinds of colonies: 'settler colonies' and 'non-settler colonies'; and to distinguish also decolonization as formal independence and membership in the international system for the colony from decolonization as integration of the colony in the parent state. In its original meaning, in Greek classical antiquity, a colony was a group of settlers who left their parent city-state to settle in an uninhabited land and form a new city-state. Cultural and sentimental links were kept with the parent state but no political domination was involved. In modern history, this form of colonization is not found in its pure form. This is so for two reasons: the parent state kept its rule over the colonists and indigenous inhabitants already occupied the land where the colonists settled. Settler colonization, in modern times, therefore, involved a double domination: that of the parent state over the settlers and that of the settlers over the indigenous inhabitants.

When the European settlers were numerous and strong in a colony, decolonization, without exception, took the form of transferring sovereignty to the settlers, not to the indigenous inhabitants. Decolonization as transfer of sovereignty came about either through the settlers using force against the parent state, as was the case for the United States and the whole of Latin America except Brazil, or through the parent state transferring sovereignty peacefully to the settlers, as was the case in Brazil, Canada, Australia and New Zealand. Decolonization in settler colonies did not put an end to the domination of the settlers over the indigenous inhabitants, however. On the contrary, once the restraint imposed by the parent state had gone, the powerful settlers were free to find a 'final solution to the native problem'. Sometimes the natives were pushed into the more inhospitable regions; often, the indigenous inhabitants had no resistance to the diseases introduced by the settlers; those who survived were put into 'reservations'. At the limit, the settlers resorted to genocide as a 'final solution'. The settlers thus became the new natives, so to speak: Americans, Australians.

The dynamic of settler colonization-decolonization overseas has radically changed the course of modern history and transformed the world. Not least because settler colonization gave birth to what became a superpower: the United States. The size and power of Russia were also enormously augmented by settler colonization. In this unique case of European settler-colonization in Asia, however, the expansion was overland, across the vast bulk of Eurasia, all the way to the Pacific. Decolonization in this case took the form of integration of the settler colony with the parent-state.

In time, the word 'colony' came to be applied to a different phenomenon altogether: the imperial rule of European states over the peoples of Asia and Africa. On the African littoral of the Indian Ocean, some European settlers were present, but, as they were even more of a minority than in South Africa, settler states were not a viable proposition. Except for the special case of Israel, there was no settler colonization on the Asiatic rim of the Indian Ocean. The 'colonies' were 'non-settler colonies'. By far the largest and the most significant of these non-settler colonies was India with its ancient and prestigious civilization. In this kind of colony, a few European administrators, soldiers, and entrepreneurs exercised political and economic power with the collaboration of indigenous elite groups. The European colonizers were not interested in just ruling in Asia and Africa; they also had an ideology of 'progress': Asia and Africa were to be transformed in the image of Europe. Through the 'civilizing mission' of Europe, modern nation-states resting on prosperous capitalism would eventually come into being in Asia and Africa. The political left was at one with the right on this: Marx was if anything even more certain of the 'progressive' role of Europe in Asia and Africa than the colonizers themselves. With decolonization, the United States and the Soviet Union took on the 'white man's burden' by vying with each other for 'developing' the new states. The term 'decolonization' came to be used for the ending of that form of rule by foreigners in Asia and Africa following the end of the Second World War.

After the New Thirty Years War (1914-1945), international politics ceased to be Europe-centred. A new configuration of world power emerged, polarized on two gigantic states, the United States and the Soviet Union, each the size of a continent extending from the Atlantic to the Pacific. The two giant states controlled the bulk of the world's power, but their rivalry in all parts of the planet was not only about power. The internal political regimes and ideologies of the superpowers were radically different. Beyond the rivalry for power, the two giants were engaged in a conflict about the kind of internal political regime the states of the world should adopt. The technology of armament produced a weapon of mass destruction, the nuclear bomb, which was qualitatively different from all other weapons. The terror of the consequences of an all out hot war fought with nuclear weapons kept the 'bellicose peace' or 'Cold War'.

The third world became a stake of the Cold War. The dialectic of the Cold War made for decolonization. The minimum objective of the policy of the United States towards the third world in the Cold War was that the new states coming out of decolonization should not adopt the model of development of its rival. If the European colonial powers refused to decolonize or delayed too long in transferring power to the westernized nationalist elites in the colonies, there was the risk of these elites being displaced by more radical elements that might then turn to the Soviet Union in the Cold War. As all the European colonial powers, condemned to mediocrity through lack of size, were allied to the United States in the Cold War, and were dependent on Washington for their security in Europe, America could use its considerable influence on them to get them to move out of their colonies. With both superpowers wooing the nationalists inside and outside the United Nations, a snow-ball syndrome of decolonization came into effect: the more colonies were decolonized the more the international political system acquired members which were for decolonization. This went on until all but the very tiniest of overseas territories became sovereign states or were integrated with their parent states.

All the new states that emerged from decolonization became embers of one and the same global international political system. Albeit that for some of the new states membership did not mean much more than a seat at the United Nations. The universalization of its membership did not alter the fundamental structure of the international political system, which it inherited from its European genesis. The anti-colonial nationalists, far from being against the land surface of the planet being compartmentalized into territorial sovereign states coexisting in anarchy, had had no greater ambition than that of membership in this international system.

The West in the Geopolitics of the Indian Ocean

Seaways, Settlers and Subsystems

The Indian Ocean, unlike the Atlantic and the Pacific, is closed on its most significant northern rim by the bulk of Eurasia, opening freely only to the south. This gives the narrow entry points to the Indian Ocean: in the north-west, the Suez-Red Sea corridor, in the east the Malacca Strait and the passages through the Indonesian islands, and the Cape route around South Africa, considerable strategic and commercial importance. These so called 'choke points' are the locations on the rim of the Indian Ocean where much weaker states can pose a threat to the sea power of the great maritime states. It is no accident that settler states were established in the offing of these crucial waterways. Two of these settler states, Australia and Israel, have proved successful and remain of outstanding importance for the West in the geopolitics of the Indian Ocean. 'Successful' in this context means irreversible settler colonization and political power in the hands of the settlers. There can be no decolonization in the sense of the physical removal of the settlers and the territory returning to the inhabitants who were there before the arrival of the Europeans. South Africa, the third state that stood guard on the critical sea routes to and from the Indian Ocean, has recently failed as a settler state.

Australia, the largest successful settler state outside the Americas, is more a continent in itself than a large island of Asia. Although it now takes an interest in the Indian Ocean, identifies itself to some extent with Asia, and has changed its immigration policy, for long it looked almost exclusively to Europe for its identity and for people, while, at the same time, being oriented to the Pacific. The history, the political regime, as well as its policies make Australia very much part of the West and a firm ally of the United States. Australia provided the United States with military facilities during the Cold War and these facilities are still available now. Middle East oil bound for Japan and the 'Asian Tigers', that are all closely linked with the West, use the sea routes between Australia and mainland Asia. The war ships of the American navy usually enter the Indian Ocean from the Pacific, through the Malacca Strait, but if need be they can go round the south of Australia. The nuclear submarines most probably take that route. In the Cold War, the Soviet war ships deployed in the Indian Ocean were based in Vladivostok and the Australian navy kept an eye on them as they went back and forth between the Pacific and the Indian Ocean. Australia has participated in the ' wars of containment ' on the mainland in Asia. A subordinate interstate system (subsystem) has not emerged, however, opposing the settler state and Indonesia or states in mainland Asia.

Even more on the periphery of the Indian Ocean than Australia, but again in the offing of another crucial entry point of that ocean, is the exceptional case of Israel. Israel is the unique case of a successful settler state that was created in the age of decolonization of non-settler colonies. Israel came into existence through the use of force and it survives through superior force. The inhabitants of the territory before the arrival of the settlers, who were displaced to make room for the settler state, have not accepted their expulsion as a 'final solution'. However, they were not capable of reversing the fact of the settler state through the use of force. The military victories of Israel, however, have not solved its security problem, for, with the acquisition of more territory, has come populations that the settler state cannot integrate. The principle on which the state of Israel is based makes it impossible to reverse the 'final solution' and accept the 'right of return' of the pre-settler inhabitants to coexist with the settlers in a secular state. On the contrary the 'right of return' is exclusively for Jews as defined by the settler state.

A subordinate interstate system developed around the native-settler conflict opposing Israel to the Arab states that identified themselves with the displaced Palestinians. By far the most important of these Arab states was Egypt. With Egypt the Arabs stood a chance, if not to win at least to avoid losing the wars in the subordinate interstate system. Without Egypt, humiliating defeats at the hands of the settler state are inevitable. The wars between Israel and the Arab states had direct geopolitical consequences for the Indian Ocean for they impacted on the Suez Canal. Indeed, between the Six Day war in 1967 and the Yom Kippur war in 1973, the Suez Canal was transformed from a waterway of global significance into a moat for the defence of the newly acquired frontier of the settler state in Sinai. Before the Six Day War the ships of Israel were banned from the use of the Suez Canal and the Arab states had made the Israeli port of Elat unusable by blocking the entry to the Gulf of Aqaba at Sharm el Sheik. The reopening and eventual enlargement of the Suez Canal to attract back the oil traffic were even more important considerations than regaining its territory of Sinai in Egypt's decision to make peace with the state of Israel.

The subordinate interstate system of the Middle East was polarized on the Cold War until Camp David and the realignment of Egypt with the United States. Moscow had helped Egypt right up to the Kippur war while the United States supported Israel. From the geostrategic point of view, the reopening of the Suez Canal had been more advantageous to the USSR than to the West. The war ships of the Soviet Union were, on the whole, smaller than those of the West and could use the Suez Canal; and that greatly reduced the distance from their base in the Black Sea to the Indian Ocean. After losing their position in Egypt the Soviets remained on the sea route by acquiring facilities in the Red Sea, in Ethiopia and in Aden on to India and linking up with the fleet from Vladivostok.

With the end of the Cold War, the United States has remained closely involved with Israel. Israel is undoubtedly a military asset for the West on the periphery of the Indian Ocean. A settler state coming into being and proving a success when most of the states on the rim of the Indian Ocean emerged from decolonization is one of the fundamental causes of instability and strife in that sector of the global political system.

South Africa is a failed settler state in the sense that the 'final solution' that the settlers adopted, Apartheid, was not acceptable to the inhabitants of non-European origins, and generally in the international system as a whole, and has been reversed. The basic cause of the non-success of the settlers was demographic: the settlers of European antecedents were always too few relative to those of non-European origins to make the settler state irreversible in the long run. Apartheid South Africa was a firm ally of the West in the Cold War. South Africa possesses a strategically significant location on the highway of the sea linking the Indian Ocean and the Atlantic. The Roaring Forties cause the ships to keep close to the coast when rounding South Africa from east to west. Apartheid South Africa monitored the traffic and provided facilities for war ships as well as the giant tankers taking oil to the West. The Very Large Crude Carriers (VLCC), 150,000 to 300,000 tons that were built after Suez was closed could not go through the Canal after it reopened. There are plans to enlarge the Suez Canal so that eventually it could accommodate even the Ultra Large Crude Carriers (ULCC) of over 300,000 tons. But the Cape Route, although a much longer route than Suez for ships bound for Mediterranean-Europe, has the advantage of ships having no tolls to pay and no restriction of size. Even if the risks posed by renewed wars between Israel and the Arab states are left out of account, the risks of pollution posed by loaded ULCCs going aground in the narrow corridor between Bab el Mandeb and Port Said, with the consequence of very high insurance premiums, are likely to keep the giant ships on the Cape route. Militarily, the Cape route has always been more secure for the dominant sea power than Suez in times of war. In a hot war, the West, with dominant sea power and the use of the South African facilities, could have denied that route to the Soviet Union and made the Suez Canal unusable to its enemy. Besides being by far the most developed capitalist economy in Africa, the minerals of South Africa were next to indispensable for the West during the Cold War. In spite of these assets, Apartheid South Africa, unlike Israel, was an embarrassment for the West that welcomed, if it did not cause, the passing away of the settler state.

A subordinate interstate system had developed around the settler native-conflict in Southern Africa during the Cold War. The subsystem of Southern Africa involved the Apartheid state on the one side, with allies for a time in Rhodesia and the Portuguese colonies, and, on the other side, Black Africa, principally what came to be called the Front Line States. With its modern military apparatus, the Apartheid state could easily hold its own in the confrontation with the Black states. To the extent that the Southern Africa subsystem polarized in the East-West Cold War the Soviet Union helped the Blacks, notably through the military intervention of Cubans. With the departure of the Cubans, and the Cold War drawing to a close, the West was less inclined to support the regime and more willing to impose economic sanctions to accelerate the changes underway inside the Apartheid state. The Apartheid system had become an impediment to the further growth of capitalism in South Africa by limiting the size of the market and the pool of trained high skilled labour that it could draw on. The abandonment of Apartheid did not mean the end of the capitalist mode of production; on the contrary, it is only if capitalism prospers that the new post-Apartheid South Africa can cope with the enormous social problem of raising the standard of living of the overwhelming Black majority. Consequently, the settlers that largely control the economy remain an essential part of the new South Africa, even though they no longer dominate the political system. With the end of Apartheid the subordinate interstate system of Southern Africa has disappeared. The new South Africa puts much less emphasis on military power than the settler state did. Indeed, South Africa is the unique nuclear weapon state that has discarded its weapons. There are signs that a Southern African region, under the leadership of South Africa, might come to replace the interstate subsystem of Southern Africa, with the blessings of the West.

Land-sea Asymmetry and Transfer of Power from Britain to the United States

The considerable importance of the Cape Route, Suez and Malacca is due not only to the access that they give to the Indian Ocean but also because, together with the Indian Ocean itself, they form part of the oceanic highway that goes right round the globe. However, besides giving high significance to the access ways of the ocean, the configuration of land and sea, with Eurasia blocking the Indian Ocean to the north, created a geopolitical asymmetry that was of enormous importance during the Cold War. The Soviet Union, occupying what MacKinder called the Heartland in the Eurasian land mass, could bring power to bear overland on to the northern rim of the Indian Ocean without necessarily taking to the sea; while the West could only intervene on land by being present at sea. The Cold War is now over, the Soviet Union has disappeared, but the asymmetry remains in that the powers inside Eurasia, and most notably China, could pose a threat overland to the states on the northern rim of the Indian Ocean. The military presence of the West in the Indian Ocean now is above all that of the United States, although two European ex-colonial powers, Britain and France, still have residual military roles there.

In the colonial days, Britain had acquired an Empire over virtually all the lands surrounding the Indian Ocean. Dominant sea power over the Indian Ocean was without doubt one of the determining factors of the British position to the East of Suez. To dominate the Indian Ocean, the

Royal Navy did not have to be physically present in force all of the time. One of the great advantages of sea power is its mobility. Before the rise of Japan, and so long as threats to the British position could only come from Europe, the British fleet could be in the Atlantic and rapidly return to the Indian Ocean in an emergency. The British Empire around the Indian Ocean, however, did not rest on sea power alone. Britain had also become a formidable land power in Asia through the Indian army. The British colonizers had organized and trained a large modern army in the Indian subcontinent. The army of the Raj not only completed the conquest of the subcontinent for Britain and 'contained' the Russian Empire to the north of the Himalayas, it also enabled Britain to extend its rule over the Middle East, over South East Asia, and further afield in East Africa. The Indian taxpayers paid for the army of the Raj, which was largely manned by Indians. With decolonization, Britain lost the Indian army and it was only a question of time before the whole of the British position, on land, East of Suez, had to go.

Decolonization on land never meant that the West was abandoning power at sea. On the contrary, with the end of colonial rule in Asia, the geopolitical situation went full circle back to the West relying primarily on sea power. This was particularly marked in the Indian Ocean. With power at sea, the West could deter a possible invasion of the land on the rim of the Indian Ocean by the Soviet Union or China, and provide reassurance to the newly decolonized states. Sea power also provided a fallback position if 'containment 'on land failed.

The United States had pressurized its European allies to decolonize in Asia. This, however, did not apply to the sea, where, on the contrary, the allies were encouraged to stay. Britain in particular, the traditional dominant sea power East of Suez, was persuaded to remain in the Indian Ocean. Washington had not succeeded in persuading London to commit land forces in Vietnam, but the UK was quite willing to stay on in the Indian Ocean provided some financial help from the USA was forthcoming. The original plan of the UK to remain present East of Suez on a severely limited budget envisaged a whole string of island bases across the Indian Ocean linking Africa with South East Asia and Australia. The Royal Air Force had put forward the scheme that the British government favoured against the Royal Navy's more costly plan to replace three ageing aircraft carriers. Britain intended to buy fifty F-111 airplanes from the United States and the striking power of these based in the islands would then be the principal means of exercising power over the Indian Ocean and of intervening on land in Asia. The financial consideration meant that the United States was involved in the British scheme from the start. Nuclear strategy in the Cold War may also have been a consideration in Washington. Polaris A-3 came on stream in 1963 and may have been deployed in the Indian Ocean at the time. There is no consensus among military experts on this. With a range of 2500 km, Polaris A-3 could reach sensitive parts of the USSR from the Arabian Sea. If Polaris were deployed in the Indian Ocean at the time - and the Soviet Union certainly reacted as if it was - then a specially equipped surface station to communicate with the Polaris nuclear submarines somewhere in the islands would have been useful, although the United States already had a communication station in North West Cape in Australia. Appropriate satellites were not available in the 1960s, and, even today, it is still useful to have surface facilities for communications.

A joint British-American team of experts surveyed the islands in 1958 and again in 1960. Initially, the islands of Aldabra, Farquhar, Agalega, Desroches, Mahe, Mauritius, Diego Garcia, Gan and Cocos were considered as suitable sites for bases, but later the British scheme of a transoceanic link of island bases had to be scaled down and then abandoned. For notwithstanding the British Prime Minister Harold Wilson's reluctance to leave the USA and China 'eyeball to eyeball', financial constraints made it impossible for the UK to keep a permanent military presence East of Suez. The UK's role was thenceforward seen as being primarily to provide the USA with secure real estate in the Indian Ocean.

A whole chain of islands was not necessary for the USA's military purposes. What were required were one or two well-located small islands capable of being transformed to accommodate substantial military facilities. Certain islands were better placed than others for military purposes, but, although location was the principal consideration it was not the only consideration, in the final decision of which islands were selected. A well-located large military base under British control already existed in the island of Gan (part of the Maldives), but was not retained. Most importantly, in the age of decolonization, it mattered for the Pentagon planners that the islands selected should have no local inhabitants and should not be under the sovereignty of a newly-independent and potentially unstable state. The UK therefore planned to decolonize its existing oceanic island dependencies, but, in the process to dismember two of them, Mauritius and Seychelles, in order to retain control over the islands selected as suitable for military use. Dismemberment and depopulation, in the age of decolonization, was seen as likely to be less problematic in Creole colonies such as Mauritius and Seychelles.

In the middle of the Indian Ocean there are three small archipelagos. The largest of the three by far is the Mascarenas archipelago composed of Mauritius, Reunion and Rodrigues; the Seychelles group is the next largest; and the third, the Chagos archipelago, is very small, but now one of its atolls, Diego Garcia, houses an American military base that has

Indian Ocean-wide significance. All of these islands were totally uninhabited when the European navigators first arrived in the Indian Ocean. The settler colonization of these previously uninhabited islands was original in that the vast majority were 'unwilling settlers', slaves from Madagascar and Africa at first, and then, somewhat more willing, indentured labourers, mostly from India. The half a million Indians who came to Mauritius dwarfed all the previous arrivals and permanently changed the ethnic composition of its population.

By the time the Indians arrived in Mauritius, the steep pyramid-like structure of Creole society was in place: at the top a small number of whites, owning most of the land, at the bottom a large black sugar proletariat, and an intermediate group in size and colour in the middle. The few Britons administering the colony grafted themselves at the top of the pyramid. The Indians had largely replaced the blacks at the bottom. Some upward mobility was possible in the rigid social structure of colonial Mauritius. Unlike the black ex-slaves, that had become largely redundant, some Indians acquired cane land and thus an economic base to finance the education of their children for jobs in the administration and in the professions. In time, a new elite emerged from the Indian labourers. Although creolized to a large extent, this new elite did not merge with the older Creole one but retained enough Indian ethnic identity for the politics of decolonization in Mauritius to be marked by 'communalism'. Britain was able to manipulate the ethnic-political division of Mauritius in decolonization in order to retain the Chagos islands for military use.

Diego Garcia

The Chagos and the other 'lesser dependencies' of Mauritius were very marginal in the sugar plantation economy. Altogether these dependencies had less than 2000 inhabitants. The white Creole elite in Mauritius knew very little, and cared even less, about the *Zilois*, as the inhabitants of the dependencies were called. The Mauritian politicians considered that the excision of small, distant islands and the expulsion of some poor, illiterate, black Creoles were a small price to pay for state power in decolonization with British support, especially as the islands were required for the purpose of enhancing the security of the West, with which Mauritius at the time identified itself. Furthermore, the British, trusted patrons of the politicians in Mauritius, had promised that the Chagos islands would be returned when the security need for them 'disappeared'. Three other small groups of islands, Aldabra, Farquhar, and Desroches were cut off from Seychelles in 1965, at the same time that the Chagos were amputated from Mauritius, and put together to form a new colony, the British Indian Ocean Territory (BIOT). However, in 1976, when

Seychelles became independent, Diego Garcia in the Chagos had finally been selected for the military base; the other three groups of islands were therefore returned to Seychelles. The large airport that had been built in Seychelles with BIOT money, and was intended to double as a military facility, was also handed over to the new state. Henceforth, BIOT comprised only the Chagos archipelago.

The dismemberment of Mauritius as a colonial entity was not unprecedented. Under the French, Mauritius had been the administrative centre for all the Creole islands. After the British conquest, Reunion had been cut off and handed back to France in 1815. In 1903 the Seychelles had been removed and made into a new colony. The difference this time, with the amputation of the Chagos, is that it took place in decolonization and violated the United Nations resolution 1514 that forbids the dismemberment of colonies. The Chagos had a resident population. The removal of the people of Chagos from their home-islands without their consent to make way for the military base in Diego Garcia was a violation of their human rights. Moreover, the British government did the dismemberment and depopulation for the military purposes of a foreign power that paid for the operation, and then lied to the British Parliament and to the United Nations to cover up this 'very sad and by no means creditable episode in British history.' The politicians in Mauritius collaborated fully in the act at the time.

The British government, with the money provided by the Americans, bought out a copra company that exploited the Chagos and deported the 1500 *Zilois* to Mauritius where they were left mainly destitute in the slums of Port Louis. A few were left in the Seychelles. In 1972 the British government gave the Mauritian government £650,000 to re-house the deported *Zilois* but the islanders never saw much of that money. What was not foreseen was that politics in Mauritius would rapidly turn against those who had done the deal over the Chagos. Soon after the independence of Mauritius in 1968, a new non-communalist and initially radical left wing political ·party, the Mouvement Militant Mauricien (MMM), rapidly gathered strength on the basis of opposition to the politicians who had collaborated in the dismemberment of the territory and the expulsion of the *Zilois*. India, whose call for help in the 1962 war with China had been used in London and Washington to rationalize the need for a military base in the Indian Ocean, now led a growing campaign against the militarization of Diego Garcia, supported by the Non-Aligned Movement and the Organization of African Unity (OAU). Under pressure from both inside and outside Mauritius successive governments in Port Louis have requested London to reopen the Chagos file on the two related issues of the *Zilois* and that of sovereignty.

On the issue of sovereignty over the Chagos, Britain has turned down all requests to reopen negotiations, and referred the Mauritians to the agreed terms of decolonization, namely: that, if the need for facilities disappeared, the Chagos islands would be returned to Mauritius. However, as an exchange of secret telegrams in November 1965 made clear, "a decision about the need to retain the islands must rest entirely with the United Kingdom Government". For good measure, those in Mauritius who had made the deal over the islands were told by their British patrons that "it would not (repeat not) be open to the Government of Mauritius to raise the matter, or press for the return of the islands on its own initiative". For internal political consumption, the Minister for Foreign Affairs of Mauritius, returning from London on one of several failed attempts to get the British government to move on the Chagos question, had this to say: "Diego belongs to Mauritius; there is no disagreement about that . . . 'Ile est a l'ile Maurice; l'usufruct est a la Grande Bretagne". There is little likelihood that the need for the military facilities will "disappear" in the near future. The Cold War has ended but the base in Diego Garcia has become even more important in the post-Cold War international system for the interventions of the United States in the Arc of Crisis. This is what is thought of in the West as a particularly unstable part of the Indian Ocean rim extending from the Horn of Africa to Afghanistan and centred on the oil-rich Gulf. The region is also a prime suspect of harbouring anti-West terrorists.

Britain is now faced with a most embarrassing situation. In 1982, the *Zilois* by then well mobilized politically, but still largely unemployed, received four million pounds from London on the condition that they gave up their demand to be returned to the Chagos. The islanders took the money but they did not cease their campaign for repatriation, and, in November 2000, the High Court in London found that the deportation of the *Zilois* had been illegal. This judgement means that the *Zilois* have the right to return to their home-islands and to receive compensation for their enforced deportation. Now Britain has recognized the *Zilois* as full British citizens. This latter move might be to weaken the claim of Mauritius to sovereignty in the Chagos, for, under the principle of self-determination, Britain might claim that the *Zilois* have freely opted to be British citizens. If the *Zilois* go back to the Chagos as British citizens, Britain can claim that BIOT is British through the free choice of its inhabitants. The British Foreign Office has commissioned a team of experts to investigate the feasibility of resettling the *Zilois* in some of the Chagos islands, although not in Diego Garcia itself. Britain, however, has treaty obligations vis-à-vis the United States and the latter flatly refuses to have the *Zilois* back in any of the islands: "[. . .] any move to settle a permanent resident population on any of the islands of the Chagos archipelago [. . .] would significantly degrade the strategic importance of a vital military asset unique in the region".

The development of a major military base on Diego Garcia rests on a treaty between the UK and the USA signed on 20 December 1966 making the islands of the British Indian Ocean Territory (BIOT) available for military purposes for fifty years, plus another twenty years extension, if required. To avoid Congress, a secret annexe provided for American finance of some fifteen million dollars for acquiring and depopulating the islands by cancelling a British debt owed for an R&D item on Polaris. The first military facility to be built was a communication station; later, new treaties with Britain on the 24 October 1972 and on 25 January 1976 greatly upgraded the facilities on the island, now styled Footprint of Freedom. According to US military sources Diego Garcia now has one of the longest runways anywhere in the world. The facilities includes: (1) for the US Navy: (a) a Navy Support Facility, which assists forward deployed units in the Indian Ocean; (b) a Naval Computer and Telecommunication Station, which also supports deployed ships; (c) Squadron Two, Maritime Prepositioning Ships with enough supplies and equipment to support over 16,000 troops for a month; (d) a Military Sealift Command Unit which provides logistic support to Squadron Two; (e) an Anti-submarine and a reconnaissance and surveillance squadron and operation support detachment; (f) Meteorology and Oceanography, Personnel Support, and Media Centre detachments; (2) US Air Force units and activities include detachments from: (a) the 613th Air Support Squadron which maintains supplies, equipment, and fuel for combat operations by up to 2000 personnel and 32 aircraft; (b) 630th Air Movement Support Squadron which provides passenger and cargo air movement services; (c) 18th Space Surveillance Squadron which performs deep space surveillance in the southern hemisphere; (d) 22nd Space Operation Squadron, one of nine remote tracking stations of the Air Force Satellite Control Network. (3) The US Army has a Care of Supplies in Storage Unit that is responsible for maintaining Army watercraft and support equipment that is part of Navy's Squadron Two.

The lagoon of the atoll has been dredged to accommodate large surface ships and submarines. The enlarged fuel storage facility can cope with thirty days' requirements of a carrier taskforce. There are reports that nuclear weapons are stored in Diego Garcia and rumours that torture has been used in the interrogation of prisoners captured in Afghanistan. Warehouses, barracks and recreational facilities have been constructed. The base works with the Rapid Deployment Force and Central Command. Diego Garcia supported operations during the 1979 crisis in Yemen and the Iranian crisis of 1979-81 involving the fall of the Shah of Iran and the seizure of the US embassy and staff. It was also prominent during

Operations Desert Shield/Desert Storm (1990-91), Desert Strike (1996), Desert Thunder (1997) and Desert Fox. During Enduring Freedom, the war against the Taliban and al-Qaida in Afghanistan, the giant B52s carrying cruise missiles and cluster bombs flew their sorties from Diego Garcia. The B-2, a heavy bomber with stealth features, also took part. B-2s flew all the way from the US, dropped their bombs in Afghanistan, then refuelled in Diego Garcia and flew all the way back to Whiteman base in Missouri. With Whiteman, Diego Garcia and Guam, B-2s can cover the whole of the planet. Two portable special shelters for B-2 bombers were erected in Diego Garcia in November 2002 and two more are to be completed in 2003. The special shelters are necessary because of the paint incorporated in the stealth feature of the B-2s.

There are about 3,000 people in Diego Garcia, of whom 1,500 are military personnel and the others are civilian workers from the Philippines and from Mauritius. The recruiting agents in Mauritius have strict orders not to recruit *Zilois*. There are no more than fifty British military; mostly Royal Marines led by a Lieutenant Commander. Their task is to see that British law and order is maintained in Diego Garcia and they also police the rest of the Chagos. The British military perform customs duties, notably seeing that drugs are not brought in. There is a British club. The Americans have a kind of amused patronizing attitude towards 'the Brits' who take their role of representatives of sovereignty rather seriously.

Each time that Diego Garcia has been used for an American intervention in one country or another on the rim of the Indian Ocean, the British government has given its approval. As the base is indispensable for effective actions, an international crisis in British-American relations could arise in the event of London and Washington being in disagreement over intervention in a particular country. This might be one of the reasons why Tony Blair, the British Prime Minister, was prepared to go against his public opinion in supporting President Bush for another war in Iraq.

The French Presence

The presence of France in the Indian Ocean is part of the presence of the West but with a difference. France, much less than Britain, does not always and everywhere positively value togetherness with the USA. France was not involved at all in the re-deployment whereby Britain transferred its role in the Indian Ocean to the United States. France keeps sizeable forces on land and sea in the Indian Ocean, but Paris always insists that these forces are independent of any military bloc.

France retained a military base in Djibouti after the independence of that strategically located enclave on the strait of Bab el Mandeb facing the island of Perim that belonged to the Soviet-supported South Yemen during the Cold War. France is a large exporter of arms to the states of the Indian Ocean, notably to the Arab states. The Arab states subsidize Djibouti and are pleased that the French keep an eye on the sea-lanes on which the arms shipments are carried. The facilities in Djibouti are used as a half-way staging post between mainland France and the south west corner of the Indian Ocean where the French presence is particularly strong.

The French presence in the south-west Indian Ocean takes two distinct forms. On the one hand, France keeps quite close post-colonial links with Madagascar and the Çomoro, and, on the other hand, the island of Reunion is an integral part of the French Republic. Reunion has very close links with Mauritius and Seychelles that are also Creole islands which were founded by France.

France had established colonial rule in Madagascar and Comoro in the nineteenth century largely because of Reunion and to compensate for the loss of its naval base in Mauritius. Mayotte proved unsuitable as a naval base, but opted to remain with France when the other three Comoro islands became independent in 1975. France eventually got a good naval base at Diego Suarez that it kept in agreement with Madagascar when the Great Island became independent in 1961. The French had to leave, however, when Madagascar later adopted a radical non-aligned position in the Cold War. France has retained sovereignty over four uninhabited tiny islets strategically located in the middle and down the whole length of the Mozambique Channel. During the Cold War it was feared in the West that Madagascar and Mozambique, that had adopted radical political regimes tilting towards the Soviet Union, would try to interfere with the super tankers in the Mozambique Channel. The United States was then quite pleased that the presence of France in Mayotte and in the islets, with two hundred Exclusive Economic Zones around them, ensured freedom of navigation without having to increase dependence on Apartheid South Africa.

The importance of the Mozambique Channel can be exaggerated; it is a convenient but not an indispensable seaway. The so-called 'outer passage' to the east of Madagascar adds only a small distance to the total from the Gulf to Europe. That route had the advantage of being easier for France to police from Reunion where a large artificial harbour has been built and where the French armed forces were redeployed when they left Madagascar.

France, unlike the other great maritime powers, wants to be accepted in the Indian Ocean on the same footing as the littoral states. The French argue that France is different from Britain and the United States in that it is not an external power with a military presence in the Ocean but that it belongs to the Creole islands. Reunion is the territorial and human expression of France's claim to be a nation-state of the Indian Ocean. The acceleration of the transformation of Reunion through *departemental-ization* was connected to the geopolitics of decolonization. De Gaulle had brought the decolonization war in Algeria to an end, decolonized Black Africa and Madagascar rapidly, and reconciled the French colonialists to the changed position of France in the world by recasting the traditional concept of *grandeur* in a European context. This did not mean the end of France's presence overseas, however. On the contrary, *grandeur*, in the changed world circumstances, called for a reinvigorated France to play a great part in Europe but not exclusively in Europe. Several European countries, in the new *Europe des nations*, and with France well to the fore, consider themselves to have a global vocation, which did not end with decolonization.

Brussels, taking its cue from Paris, has encouraged the drawing together of the islands to form the Indian Ocean Islands Commission (IOC). This brings together Reunion, which, as a part of France, is an integral part of the EU, and the other islands that are ACP states. The international legitimacy of Reunion as a complete part of France is thereby strengthened. When the islands recognized Reunion as France they recognized France as a littoral state of the Indian Ocean. France, in the age of decolonization, wants to be accepted as a nation-state of the Indian Ocean primarily because of the status of Reunion: the island must be seen, not as a colony, but as an integral part of France. Beyond the status of Reunion, however, France wants to be present in the Indian Ocean as part of its ambition for an independent role in world politics.

The 'global vocation' of France, as it manifest itself in the Indian Ocean, is not always welcomed with the same enthusiasm by all the littoral states of that Ocean. The Reunion as-an-integral-part-of-France position has never been challenged at the United Nations, where France is a permanent member of the Security Council. France takes the view that, even more than Hawaii is an integral part of the United States, Reunion is French by its history and by the self-determination of the people in democratic institutions. In the African context, Colonel Gaddafi raised the issue of Reunion once at a meeting of the OAU but France had no difficulty in persuading its numerous clients in Africa to quash the Libyan move. Over the Indian Ocean as a Zone of Peace (IOZP) project during the Cold War, France's argument that, as a nation-state of the Indian Ocean, its presence did not violate the Zone of Peace, was not convincing for all the littoral states. More recently, France's application for membership of IOR-ARC was vetoed by India. Paris was most disappointed that Mauritius, in particular, had not supported its membership. However, France joined IOR-ARC as a Dialogue Member in 2001.

Considering the importance that 'communalism' has in the internal politics of Mauritius, Port Louis cannot afford to ignore the position of India on issues concerning the Indian Ocean. Nevertheless, Port Louis and Mahe have usually responded very well to the intense many-sided diplomacy of France in the islands. Dependence on France is perceived as not entirely one-sided: France, it is said, depends on the Creole islands to some extent for acceptance as one of them, as a littoral state of the Indian Ocean. Moreover, France attaches a lot of importance to the *rayonnement* of the French language in the world and it needs the Creoles as allies in an Indian Ocean surrounded by countries where English is the growing language.

India: Betwixt Land and Sea

The Land

The Indian Subcontinent as well as the Indian Ocean are called after India. The enormous diamond shape mass of India occupies two thirds of the land mass of the Subcontinent, which is a natural physical-geographical entity. While India forms a compact hunk of territory at the centre, the other states of the Subcontinent are at the periphery with unfavourable shapes and frontiers, widely separated from each other by the bulk of India which is however contiguous with them all. The great rivers of the Subcontinent flow across the territory of more than one state but they all go through India. The landlocked states of the Subcontinent are dependent on India for access to the sea. Demography, civilization and culture tend to reinforce physical geography in giving India a 'natural' hegemony over the Subcontinent. India has three quarters of the total population of the Subcontinent. All the present day states of the Subcontinent have at one time or another been part of an Indian entity. The religions, the ethnies, which are present in the other states, are also found inside India. If it is possible to speak of a civilization of the Subcontinent as a whole, then that civilization is Indian. Even if all the other states of the Subcontinent were to unite they could not redress the balance of power with India. Yet India, in contrast to the United States in North America, does not have political hegemony in the Subcontinent. Nor is there an effective Indian variant of the Monroe Doctrine. India has not been able to prevent the interventions of external powers in the Subcontinent that enable much smaller states to refuse India's predominance. The Subcontinent remains an interstate system and an interstate system that developed around the India-Pakistan conflict, a sequel of the decolonization of the Subcontinent. In turn, the interstate system of the Subcontinent became involved in the global East-West conflict of the Cold War and with the Sino-Soviet conflict after the split in the Communist camp.

The formidable mountain barrier has never prevented the Subcontinent from being invaded time and again from the north. Under colonial rule, the British Raj had trained a large Indian army and established a system of buffer-states to 'contain' the overland threats to the Subcontinent. The nationalist leaders in India, however, had consistently criticized the military policy of the Raj. It was a policy, they said, for the benefit of the British imperialists and not for that of India. Indian resources were squandered in militaristic pursuits that antagonized neighbours with whom India had no real quarrels. Although at independence India inherited the lion's share of the huge military apparatus of the Raj – but, significantly, only a small navy – the nationalist leaders were not at all inclined to follow in the British footsteps, at least not as far as the military dimension was concerned. The priorities of Nehru were elsewhere, in economic and social developments.

Newly-independent India perceived no threats to its security from the Soviet Union. Furthermore, so long as China was a close ally of Moscow, then India did not feel insecure towards that state either. The defence budget was drastically cut. To join a military pact aimed at the USSR and China would just bring the Cold War into what otherwise would be a 'zone of peace'. Yet Nehru's India had an ambition to play a great role in world politics. The role that Nehru wanted for India was innovative in geopolitical terms: the countries of the northern rim of the Indian Ocean could be a neutral zone in the conflict between the power dominating the land and the power dominating the sea. Moreover, by keeping out of the military alliances of the Cold War, India would be well placed to act as a global arbiter for the sake of world peace. In occupying the moral high ground and leading the Non-Aligned Movement, India would have more influence in the world than by being a member of a Cold War alliance.

India, however, in spite of its democratic regime and its many affinities with the West, 'tilted' its non-alignment towards the Soviet Union, the superpower of the land, less through fear of the United States, the power dominant at sea, but for security vis-a-vis China after the Sino-Soviet split. China became a threat to India not only because of the unresolved border conflict but because India became for China an indirect way of fighting the ideological war against the Soviet Union. After the humiliating defeat in the border war with China in 1962, India's defence priority more than ever had to be on land. China intervened in the interstate system of the Subcontinent by supplying Pakistan with technologies for advanced weapons. Thus, Pakistan, already armed by the West as part of the East-West Cold War, was also strengthened by China.

These external interventions enabled Pakistan, an intrinsically much weaker state than India, to keep up with its powerful rival in the subsystem of the Subcontinent. The smaller states of the Subcontinent tried to remain 'non-aligned' in the India-Pakistan conflict while attempting to play on the rivalry to enhance their independence. Nepal and Bhutan also felt that the rivalry between India and China could give them room to manoeuvre. Sri Lanka was well placed to flirt with the West in the Indian Ocean.

The Sea

The central position of the Subcontinent in the middle of the monsoon sector of the Indian Ocean placed it on the routes of the oceanic navigators long before people were able to cross the Atlantic and the Pacific. The Great Peninsula penetrates the Indian Ocean from the Tropic of Cancer to near the equator, dividing the northern part of the Ocean into the two basins: the Arabian Sea and the Bay of Bengal. It is that northern part of the Indian Ocean that can really be called 'Indian'. In the pre-colonial era Indian civilization had been carried by sailors and scholars across the Bay of Bengal to East Asia and across the Arabian Sea as far as the coast of Africa. The Asian navigators, however, seldom went south of the equator; that remained largely unexplored until the arrival of the Europeans.

In the Cold War, the activities of the West in the India Ocean, more so than those of the USSR, were perceived by the littoral states as the opening of a new front that was a threat to the non-alignment that Nehru had advocated. India, although it had reservations, supported an initiative of Sri Lanka to make the Indian Ocean into a Zone of Peace (IOZP) from which the military forces of the great powers in the Cold War would be excluded. Nothing much came of the IOZP – although a UN resolution to that effect was passed in 1971. The littoral states could not agree on the extent of the coverage of IOZP: should it cover just the body of water? Or should IOZP cover the land around the rim of the ocean as well? India, for one, was against the extension of IOZP to the land of the littoral states, not least because this would have foreclosed the option of acquiring nuclear weapons. Pakistan, on the other hand, was keen that IOZP should extend to the land thus depriving India of the enormous military advantage it had in the Subcontinent. The extension of IOZP to the land was not reconcilable with the military alliance of Australia with the United States and with the position of Israel. However, even if the IOZP had been limited to the sea, the great naval powers would not have agreed to it because of the overwhelming importance of the sea-based nuclear weapons for the credibility of mutual deterrence in the Cold War. The

Indian Ocean is only a part of the global ocean and to withdraw the Indian Ocean from that global body of sea-water would have impeded the global mobility of sea power thus affecting the credibility of mutual deterrence that was essential for crises in the Cold War not triggering a nuclear war.

It has been said that the Soviet navy in the Indian Ocean from 1968 onwards posed a threat to the tankers carrying oil to the West and to Japan. However, if the Soviet intention had been to attack the tankers the Indian Ocean was the least likely stage on their routes to do so. Much more likely places to attack the tankers were near their arrival points in the Atlantic and the Pacific, where Soviet submarines were in force. It was much less the presence of the Soviets at sea than a possible threat on land in the Middle East, the source of the oil, that led the United States to step-up its presence in the Indian Ocean. The Tehran hostage crisis of 1979, the Kabul coup of December 1979, interpreted as the opening move of a Soviet drive to the 'warm waters ' were met by the Carter Doctrine, the Rapid Deployment Force and finally the setting up of the Central Command in 1990. By then the Soviet Union was ceasing to exist and had withdrawn from Afghanistan. In the mean time the West gave massive help to Pakistan in 1981 and this was renewed in 1986. The facilities in Diego Garcia were considerably extended after 1976.

Nationalist writers have stressed that all the invaders of India by land have been absorbed into India's composite civilization, while the one invasion from the sea, the European colonial invasion, had put India in chains. Yet the Indian nationalist elite that inherited power in decolonization, notwithstanding non-alignment, had welcomed Britain remaining present at sea in the Indian Ocean after independence. Under the Raj, the land power of India was developed but the British had not trained Indians for a great role at sea. The rhetoric of threats from the sea resurfaced when Henry Kissinger sent the *Enterprise* task force in the Bay of Bengal, when the Indian army was 'liberating' Bangladesh. The American move was perceived in New Delhi as an attempt at intimidation and an affront. India has put more resources into its naval defences since, and has even acquired some elements of a Blue Water navy. The Indian naval build up caused some concern around the Indian Ocean, notably in Australia. Modern naval armaments, however, demand a great deal of capital investment and are very expensive for a developing country. A credible high seas fleet capable of challenging the United States in the Indian Ocean is beyond the means of India, at least for the time being. Besides, the United States had supported the decolonization of the Subcontinent and a democratic India, albeit non-aligned, was an asset against Communism. It was international politics within the Subcontinent, the obsession with Pakistan, and the willingness of Islamabad to join Western military pacts in order to get modern weapons to 'redress the balance' with its large neighbour that really fuelled the rhetoric of the threat from the sea.

The end of the Cold War has substantially modified the geopolitical context. Non-alignment has become meaningless and India has lost the role of arbiter that it wanted to play in the global conflict. More significantly, the security of India is not any better in the post-Cold War international system. On the side of the land, the Red Army is not there any more to provide a counterweight to China and reassurance in the contingency of renewed threats on the land borders of India. India has tested nuclear devices and the means of delivering them over long distances are improving. However, India cannot yet match China at the conventional and nuclear levels. Security wise, the crucial balance of power for India now is with China and not with Pakistan, which, with or without nuclear weapons, can never be in the same league. On the side of the sea, even with a more powerful fleet in the Indian Ocean, India, at best, would only be able to deny the United States the full advantages of commanding the sea around India. To repeat, the Indian Ocean is a part of the global ocean and sea power is mobile. There can be no real reversal of the Vasco da Gama epoch in the foreseeable future, for that would call for India's sea power not only to command the Indian Ocean but also being able to challenge the West all over the global ocean. However, there is now a rapprochement of India with the United States in the Indian Ocean, perhaps as a reassurance against China that is showing signs of wanting to have a military presence in the Indian Ocean. India is moving so far away from Nehru's India that some authors have argued it calls for 're-imagining'. The Indian economy is opening up to the world market; the Indian navy participates in joint exercises with the US navy in the Indian Ocean; not one occasion is missed for stressing how much in common the 'largest democracy in the world' has with the United States. 'War on terrorism' is a further factor making for togetherness between Washington and New Delhi since September the 11th.

India, as the champion of the third world and non-alignment in the Cold War, had always supported the natives in the interstate subsystems that had developed around the native-settler conflicts in the offing of the sea accesses to the Indian Ocean. Gandhi's first political struggle was against Apartheid and India has welcomed the arrival of the new South Africa. The recent rapprochement of India with Israel marks a change in the generally pro-Arab position that was initiated by Nehru's close relations with Nasser. Tel Aviv provides New Delhi with a new source of advanced weapons and know-how in the fight against terrorism, but this cannot fail but to antagonize the Arab states from which India imports oil and exports 'guest workers'. India, even when it was still under colonial rule, had always protested against the White Australia's immigration law that kept Indians out and has welcomed the change of policy in Canberra. Now, Indians with high skills can 'brain drain' themselves to Australia. The Indian diaspora in the lands around the Indian Ocean and beyond can be a positive asset for the external policy of India but also for its internal development, as the Prime Minister stressed at the recent *Pravasi Bharatiya Divas*.

Some Conclusions

The place of the Indian Ocean as a body of sea-water in global politics is quite distinct from that of the land rim of that ocean. The Indian Ocean as a body of sea-water is a part of the global ocean that has been dominated by the West since the Age of Discoveries; and it remains so today more than ever with the United States as the only superpower in the global political system. The states in the land rim of the Indian Ocean ceased to figure in the struggles for hegemony at sea early on after the arrival of the Europeans. Decolonization in the land rim of the Indian Ocean changed the configuration of power on land but not in the Indian Ocean as a body of seawater. Relinquishing power on land did not signify that the West was giving up power at sea. On the contrary, with decolonization on land, power at sea became more, not less, important for the West.

The transfer of hegemony in the Indian Ocean from Britain to the United States involved the dismemberment in decolonization of two small Creole colonies in order to provide a safe territory for America's pivotal military base in the Indian Ocean. Diego Garcia had some value in meeting the Soviet challenge at sea during the Cold War but the real significance of the mid-oceanic base was, and remains, for the United States to project power on the land rim of the Indian Ocean and beyond on the heartland of Eurasia. Now that the Cold War is over, the power of the West is projected from the Indian Ocean primarily on the so-called Arc of Crisis, that is that part of the land rim that extends from the Horn of Africa to Afghanistan and is centred on the oil-rich Gulf. Diego Garcia is one of the factors that accounts for the particularly close coincidence between British and American policies regarding war and peace in the Arc of Crisis. Britain provides the BIOT to the United States and London has approved all the interventions from the base in Diego Garcia. Considering the indispensability of the base for the projection of Western power in the land rim of the Indian Ocean, it would have been difficult for Britain to object to an American intervention, for example, to remove the regime in power in Iraq without causing a rift in Anglo- American relations.

Britain has contrived sovereignty in the Chagos, but, so far, it has made no claim that the islands are an integral part of its national territory in the same sense that, say, the Channel Islands are. In fact, by reiterating that the Chagos would be returned to Mauritius when the military needs for them have "disappeared" London seems to foreclose such an interpretation of its position. France, on the other hand, claims that it is not an external power with military and other interests in the Indian Ocean, but that it belongs as a littoral state because a part of its national territory, Reunion, is in the Indian Ocean and that the people of the island are as French as those of Paris. Reunion is an end in itself, but it is also a means of France's 'global vocation' to be present in the Indian Ocean as in the Atlantic and the Pacific.

The land rim of the Indian Ocean from the southern tip of Africa right round to the Middle East, taking in the Red Sea and the Gulf, on to the Indian Subcontinent, the Indonesian islands, and ending in Australia, does not in any sense form a region. Rather than forming a region, the land rim of the Indian Ocean breaks down into a number of interstate subordinate systems. Of these interstate subsystems, those that are on the seaways connecting the Indian Ocean to the rest of the global ocean, and have opposed settlers to natives, have been particularly significant in the geopolitics of the Indian Ocean. One of these, the Southern African interstate subsystem, has disappeared with the end of Apartheid. If the Middle East interstate subsystem no longer opposes all the Arab states to Israel, the native-settler conflict nonetheless remains of outstanding importance, affecting a sector of the Indian Ocean rim that contains the principal oil reserves of the world and one of its major seaways.

India has given its name to the Indian Subcontinent as well as to the Indian Ocean. Although in a geographical, historical and cultural sense the Subcontinent can be said to be 'Indian', the Subcontinent is not 'Indian' in the political sense today. The Subcontinent remains an interstate subordinate system. For countless centuries, monsoon navigators have linked the shores of the northern part of the Indian Ocean and Indian culture has influenced the lands across the Bay of Bengal and the Arabian Sea, but the Indian Ocean as a part of the global ocean remains under the hegemony of the West. Were the other states that share the Subcontinent with India to acknowledge and welcome a leading role for New Delhi, instead of calling in external powers to match it, this would greatly help India to reach a position of equilibrium with China and could enhance its position in the Indian Ocean.

References

Albrow, Martin (1966), *The Global Age* (London: Polity Press).
Anderson, Benedict (1994), *Imagined Communities* (London:Verso).
Aron, Raymond (1962), *Paix et Guerre entre les Nations* (Paris: Calman-Levy).

Botbol, Maurice, and Others, eds. (Since 1981), *La Lettre de l'Ocean Indien* (Paris: Indigo). (Weekly newsletter, which is very useful for following events in the islands).

Benedict, Burton (1965), *Mauritius, the Problems of a Plural Society* (London: Pall Mall).

Bunge, Frederica M. ed. (1983), *Indian Ocean, Five Island Countries* (Washington DC, The American University).

Chalian, G. and Rageau, J-P (1985), *Strategic Atlas, World Geopolitics* (London: Penguin Books).

Chaturvedi, Sanjay, (2000) 'Representing post-colonial India: inclusive/exclusive geopolitical imaginations' in Dodds, K and Atkinson, D. ed., *Geopolitical Traditions, A Century of Geopolitical Thought* (London: Routledge).

Chaudenson, Robert (1992), *Des Iles, Des Hommes, Des Langues* (Paris: l'Harmattan).

CERSOI, eds. (Yearly since 1974), *Annuaire des pays de l'Ocean Indien*, Aix en Provence, Presses de l'Universite'de Aix en Provence (provides first class analyses of the Créole islands since 1974, as well as regular bibliographies, chronologies, and other documentation.)

Chaudhri, K.N, (1985), *Trade and Civilization in the Indian Ocean, an economic history from the rise of Islam to 1750* (Cambridge: CUP).

Cipolla, Carlo (1962), *Guns and Sails in the Early Phase of European Expansion 1400-1700* (London: Collins).

Corbridge, S. and Harriss, J. (2000), *Reinventing India* (London: Polity).

Cottrell, A. J. and Burrell, R. M., eds. (1972), *The Indian Ocean, its Political, Economics and Military Importance*, (New York: Praeger).

Coutau-Begarie, H. (1993), *Géostratégie de l'Océan Indien* (Paris: Economica).

Darby, P. (1973), *British Defence Policy East of Suez* (London: OUP).

Darwin, J. (1983), *Britain and Decolonization* (London: Macmillan).

Dowdy, W. L. and Trodd, R. B., eds. (1985), *The Indian Ocean: Perspectives on a Strategic Arena* (Durham NC: Duke UP).

Ferro, Marc (1997), *Colonization, a Global History* (London: Routledge).

Fukuyama, Francis (1992), *The End of History and the Last Man* (New York: Free Press).

Gillard, David (1977), *The Struggle for Asia 1828-1914* (London: Methuen).

Gottmann, (1952), *La Politique des Etats et leur Géographie* (Paris: Armand Colin).

Graham, Gerald S. (1965), *The Politics of Naval Supremacy* (Cambridge: CUP).

Graham, Gerald S. (1967), *Great Britain in the Indian Ocean* (Oxford: Clarendon).

Gray, C. S. and Sloan, G. (1999), *Geopolitics, Geography and Strategy* (London: Frank Cass).

Hall, Richard (1996), *Empires of the Monsoon: A History of the Indian Ocean and its Invaders* (London: Harper Collins).

Harrison, Selig and Subrahmanyam, K. (1989), *Superpower Rivalry in the Indian Ocean, Indian and American Perspectives* (Oxford: OUP).

Herodote (1993) *L'Inde et la question nationale* (Paris: Hérodote).

Houbert, Jean (1983), "Décolonisation en pays créole: Maurice et La Reunion", *Politique Africaine*, Paris, 10 (juin 1983), pp. 78-96.

Houbert, Jean (1985) "Settlers and Seaways in a Decolonized World", *The Journal of Modern African Studies*, Cambridge, 23, 1 (1985), pp. 1-29.

Houbert, Jean (1986), "France in the Indian Ocean: Decolonizing without Disengaging", *The Round Table*, London, 298 (1986), pp. 145-166.

Houbert, Jean (1992), "The Indian Ocean Creole Islands, Geopolitics and Decolonization", *The Journal of Modern African Studies*, Cambridge, 30, 3, (1992), pp. 465-484.

Houbert, Jean (1997), "Russia in the Geopolitics of Settler Colonization and Decolonization", *The Round Table*, London, 344 (1997), pp. 549-561

Jackson, Robert (1990), *Quasi States: Sovereignty, International Relations and the Third World* (Cambridge: CUP).

Jones, Archer (1989), *The Art of War in the Western World* (Oxford: OUP).

Jones, E. L (1985), *The European Miracle: Environment, Economics and Geopolitics in the History of Europe and Asia* (Cambridge: CUP).

Kennedy, Paul (1988), *The Rise and Fall of the Great Powers* (London: Fontana).

Lemon, A. and Pollock, N., eds. (1980), *Studies in Overseas Settlement and Population* (London: Longman).

Leymari, Philippe (1981), *Océan Indien, le nouveau Coeur du monde* (Paris: Karthala).

Maestri, Edmont (1994), *Les Iles du sud-ouest de l'Océan Indien et la France de 1815 a nos jours* (Paris: l'Harmattan).

Mahan, A.T. (1900), *The Problem of Asia and its effects upon International Policies* (Boston, Mass.: Little Brown).

Mahan, A. T. (1957 reprint), *The Influence of Sea Power Upon History 1660-1783* (New York: Sagamore Press).

Marimootoo, Henri (1997), 'Diego Files': a series of ten articles based on recently declassified documents at the Public Record Office in London, Week-End, Port Louis, 25 May-27 July 1997.

Martinez, Emile (1988), *Le département francais de la Réunion et la coopération internationale dans l'Océan Indien* (Paris: l'Harmattan).

Marx and Engels (1976 reprint), *On Colonialism: reprints of articles from the New York Daily Tribune* (London: Lawrence & Wishart).

Maurice, P. and Gohin, O. (1993), *Les Relations internationales dans l'Océan* Indien,(La Réunion Université).

Mazeran, Helene (1994), *Géopolitique de l'Océan Indien* (Paris: CHEAM).

Mauritius Legislative Assembly (1983), *Report of the Select Committee on the Excision of the Chagos Archipelago* (Port Louis).

Nehru, Jawaharlal (1989), *The Discovery of India* (Oxford: OUP).

Northcote-Parkinson, C. (1954), *War in the Eastern Seas* (London: Allan and Unwin).

Panikkar, K. M. (1971 reprint), *India and the Indian Ocean* (New Delhi: Allen and Unwin).

Parry, J. C. (1979), *Europe and the Wider World 1415-1715* (London: Hutchinson).

Pochoy, Michel, (1991), *Le Pakistan, L'Ocean Indien et la France* (Paris: FEDN).

Scammell, C. V. (1981), *The World Encompassed: the First European Maritime Empires c 800-1650* (London: Methuen).

Scott, Robert (1961), *Limuria, the Lesser Dependencies of Mauritius* (London: Greenwood).

Sick, Gary (1983), "The evolution of US strategy toward the Indian Ocean and the Persian Gulf", in Rubinstein, A. Z., ed., *The Great Game* (New York: Praeger).

Strang, David (1991), "Global patterns of decolonization 1500-1987", *International Studies Quarterly*, 35 (1991), pp. 429-454.

Toussaint, Auguste (1966), *History of the Indian Ocean* (London: Routledge).

Toussaint, Auguste (1972), *Histoire des îles Mascareignes* (Paris: Berger-Levrault).

Vigarie, André (1995), *La mer et la géostratégie des nations* (Paris: Economica).

Warren, Bill (1980), *Imperialism: Pioneer of Capitalism* (London: Verso).

Watson, Adam (1992), *The Evolution of International Society* (London: Routledge).

CHAPTER 4

Emergence of a New Geopolitical Era in the Indian Ocean: Characteristics, Issues and Limitations of the Indianoceanic Order

Christian Bouchard

Before the European projection that began in the late 15th Century, Indian Ocean populations were bonded by large-scale maritime trading systems and outside influences were minimal. This Pre-Gamian order (referring to the time before Vasco de Gama's trip to the Indies in 1497-98), characterized by regional self-sufficiency and autonomy, was then slowly replaced by the colonial order, characterized by European economic and political control, which culminated in the second half of the 19th and the first half of the 20th Centuries. Since the Second World War, the Indian Ocean Region has seen tremendous geopolitical change as almost every littoral state gained its independence, and then saw its international relations subordinated by Cold War constraints, and eventually established regional cooperation with its neighbours. In a very short period of time, emancipation from formal foreign control had become a very significant and profound movement.

As a result, the region has now entered into a new geopolitical era. The Indianoceanic order, as we propose to call it, is articulated around five main characteristics, which are:

1). The great political, cultural and economic heterogeneity of the region;

2). A fragmentation into well-affirmed sub-regional systems, where regional cooperation and economic integration are organized and the geopolitical equilibrium is constructed;

3). An emergent Indianoceanic regionalism, which is original in nature and is now formalized by the *Indian Ocean Rim Association for Regional Cooperation* (IOR-ARC);

4). A subordination to large foreign powers, especially to those of the industrial Triad (United States, European Union and Japan) which have a tremendous influence on the region, even if they do not exercise complete control of it; and,

5). The very importance of the Indian Ocean itself, as it represents both the major link of the region and its main door to the rest of the world.

Several questions about this new geopolitical order remain to be answered. Among others, if the emancipation trend from foreign influences is to continue, this means that stronger regional economic relations and political cooperation are to be developed. Eventually, in this context, peace, security, military and strategic concerns would have to be back on the forefront of the Indianoceanic agenda.

Introduction

Since the Second World War, a new geopolitical order is emerging in the Indianoceanic Region. Through the decolonization process which began on the shores of the Indian Ocean as early as 1947 with the Independence of India and Pakistan, the people of the region have regained one after the other the political control over their respective territories. In the 1960s, the old colonial order was replaced by a new order which we will refer to as the Indianoceanic order. However, at the same time, the emancipation of the new born states was seriously slowed down by the projection into the Indian Ocean area of the great ideological, economic and strategic rivalry that opposed the two large superpowers of the time, namely the Soviet Union and United States. In the Cold War period, almost all of the Indianoceanic states have been compelled to choose a side, even if they claimed to be non-aligned. This situation has profoundly impacted upon the geopolitical equilibrium as neighbours sought the support of one or the other superpower as balancer of power. In this context, multilateral economic and political cooperation among all the states in the same geographical area was almost impossible to achieve.

At the end of the 1980s and the beginning of the 1990s, the collapse of the communist bloc, the disengagement of the Soviets from the international scene and the dissolution of the USSR have dramatically changed the world's geopolitical system. In the Indianoceanic Region, the end of the Cold War found expression in a much greater autonomy for states in regard to their international relations, allowing them to develop ties with all neighbours as well as with other more distant states. At the same time, regionalism was fostered by the emergence of regional powers and their leadership in the establishment or the renewal of cooperative dynamics among neighbouring states. Finally relieved of the constraints of the East-West confrontation, the emancipation of the Indianoceanic states could continue and take on a new dimension. Having regained political control over their territories, it is now time for the Indianoceanic people to regain control of their regional area at both the neighbourhood and the larger Indian Ocean scales. In the former case, this enabled economic integration and sectorial cooperation to develop and revitalize the formal international associations of regional vocation. In the latter case, a new and original dynamic of large-scale regionalism has emerged in the second half of the 1990s, a process that is largely related to the conjunction of three fundamental developments which are: India's economic opening, the reintegration of South Africa as a normal member of the international community and the fact that Australia became aware of its Indian Ocean interests.

However, the return of some Indianoceanic states (especially the largest regional powers that are India, Australia and South Africa) to the forefront of the regional geopolitical system does not mean that the region is no longer subordinated to the global system, and, thus, to the large world powers (especially those of the industrial Triad) as the latter continue to dominate the world system. Therefore, in contrast to the Pre-Gamian and colonial orders, respectively dominated by bordering peoples and European colonial powers (especially the United Kingdom as, from 1815 to the mid-twentieth century, the Indian Ocean was nothing less than a "British Lake"), the Indianoceanic order is more complex as both local states and foreign large powers share the control of Indian Ocean affairs. Today, it seems possible to identify the main characteristics of this new Indianoceanic order, an appellation that refers to the regional solidarities that are actually developing and which could play a significant role in the future of the region.

Definition of the Indianoceanic Region

Proposing a definition for the Indianoceanic Region presupposes that we have a clear purpose for this definition, and, secondly, that we can identify the criteria that are the most meaningful in this context. If it is true that delimiting a region is a very subjective exercise, then this does not necessarily imply that the result is futile as, when done properly, it should be representative of a certain reality. If common characteristics, actions and interests are to be found among neighbours of a specific area, then it should be possible to identify a regional system of interactions linking more or less closely the implicated states. As the purpose of our work is specifically geopolitical, then the factors determining our regionalization will be essentially geopolitical.

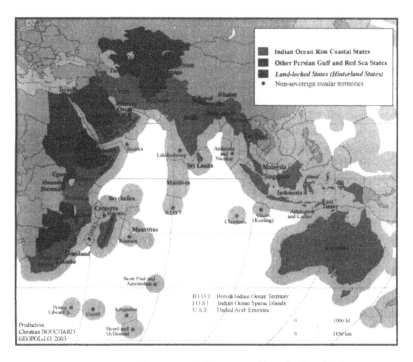

Figure 4.1. Indianoceanic States and Insular Territories.

For instance, in the context of the Declaration of the Indian Ocean as a Zone of Peace (IOZP) which was the object of UN resolution 2832(XXVI) of 16 December 1971, the extent of the Indian Ocean Region was defined in 1979 as the Indian Ocean itself, its natural extensions [1], the islands thereon, the ocean floor subjacent thereto, the littoral and hinterland states and the air space above [2]. This definition can be said to be broad, as it considers the coastal states of the Persian Gulf [3] and of the Red Sea, as well as the land-locked states which transit to and from the sea but would mostly be oriented to the Indian Ocean (Figure 4.1). In the wake of recent geopolitical developments in Central Asia and the plans to export some of its oil via the Indian Ocean, the five ex-Soviet Republics of the area can be considered to be part of the regional hinterland. As Indonesia is without any doubt an Indian Ocean coastal state, and, given the archipelagic nature of its territory, then its waters should be included as part of the Indian Ocean Region, and East Timor, located between Indonesia and Australia, could therefore also be considered as an Indianoceanic state, even if the Timor Sea is often associated with the Pacific Ocean. Finally, as the 60th South parallel marks the limit of the area covered by the Antarctic Treaty, then it appears very logical to consider this parallel as the southern limit of the Indianoceanic Region. South of this parallel, economic and environmental issues as well as the political and strategic context clearly form a distinct geopolitical area.

A more narrow definition would consider only the littoral states and eventually only those that possess maritime coasts or port facilities on the main body of the Indian Ocean, therefore excluding some coastal states of the Persian Gulf (Bahrain, Qatar, Iraq and Saudi Arabia) and the Red Sea (Egypt, Eritrea, Israel, Jordan and Sudan) as well as the land-locked states (Afghanistan, Botswana, Bhutan, Burundi, Ethiopia, Lesotho, Malawi, Nepal, Rwanda, Swaziland, Uganda, Zambia and Zimbabwe). For example, this definition corresponds to the understanding of the Indian Ocean Rim as it is actually used by the *Indian Ocean Rim Association for Regional Cooperation* (IOR-ARC).

Making a choice between the broad and the narrow definition makes a lot of difference in regard to the geopolitical characteristics of the region. For example, the former implies that Israel and Jordan, Kuwait and Iraq, Botswana and Zimbabwe, Rwanda and Burundi, Bhutan and Nepal, as well as Saudi Arabia and others would be included in the Indianoceanic Region. In contrast, if one accepts France (for Reunion, Mayotte, the French Indian Ocean Sparse Islands, Crozet, Saint-Paul and Amsterdam and Kerguelen) and the United Kingdom (for the British Indian Ocean Territory) as Indian Ocean coastal states, then the narrow definition will only include 28 states, which is only half of the 56 states which would be included within the broad definition (Appendix 4.1).

It appears to us that if the Indian Ocean is to be the core element of this region, then maritime affairs should be of primary concern in defining the area of the region. At the very least, all the Indianoceanic states will share the same concern of using the Indian Ocean as their unique maritime connection or one of their maritime connections to the rest of the world. On this basis, Indian Ocean related land-locked states would be considered as part of the region. Regarding the case of the Persian Gulf, as it is a natural extension of the Indian Ocean, which also represents its unique orientation to the rest of the world, it would make a lot of sense to include its coastal states in the Indianoceanic Region. To help us on this issue, we can also consider that, by adding Persian Gulf coastal states to the region, it is therefore possible to consider more globally the very significant oil traffic from its sources to the chokepoints by which oil tankers leave the Indian Ocean.

Concerning the Red Sea, it is another natural extension of the Indian Ocean, which links it to the Mediterranean Sea by the way of the Bab el Mandeb and the Suez Canal. Not considering Egypt where the canal is located as an Indianoceanic state while at the same time considering Singapore at the southeastern end of the Malacca Strait to be an Indianoceanic state does not appear to be very consistent. At the same time, Sudan shares a maritime common zone of exploitation with Saudi Arabia, and Eritrea is one of the transit states for Ethiopia, just like Djibouti and Somalia. Thus, at least for Egypt, Sudan and Eritrea, on the basis of maritime affairs, it seems to be logical to include them into the group of Indianoceanic states. This leaves us with the question of Israel and Jordan that both possess a port at the southern end of the Gulf of Aqaba. In practice, this would be enough to make them Indianoceanic states in terms of the broad definition. Excluding them would mean that they would have no say in the discussion on maritime affairs even if they are in fact very much interested in these as Aqaba is the unique Jordan port and the Israeli navy is operating in the Red Sea. One should also remember that, following the Six Days War, Israel enforced the closure of the Suez Canal between 1967 and 1975. Thus, for these reasons, it appears that Israel and Jordan should also be considered as Indianoceanic States.

In terms of the geopolitical issues commonly shared by Indian Ocean states, maritime issues are more relevant than any other considerations and for that reason the broad definition should be used (Figure 4.1 and Appendix 4.1). This interpretation does not mean that a formal regional association should necessarily include all of these 56 states, especially if its purposes are mainly commercial. However, eventually, if this Association is to seriously address maritime affairs as well as peace and security issues, then it would have to be a forum with a very broad membership.

The Great Heterogeneity of the Region

The Indianoceanic Region covers 105.6 million sq km, of which the Indian Ocean itself accounts for 68.6 million sq km (CIA, 2003). Together, the continental and insular lands of the 56 Indianoceanic states cover 37 million sq km, hence representing 35% of the total regional area, and are inhabited by some 2.46 billion people (PRB, 2003). The first characteristic of the region is its very great heterogeneity in terms of territorial, demographic, political, ethnic, cultural, economic and military matters. While the unequal resources distribution is an indisputable permanence, the great disparities of development among Indianoceanic societies will also remain for a long time as a very significant factor to consider, more especially as in the next decades population growth will continue to be high in those states that have, until now, possessed lower levels of human development. Clearly, the fact that the political (states, territories), ethno-cultural (ethno-cultural groups) and economic (activities and potentialities) maps do not match one another represents a fundamental element of the regional geopolitical framework.

In regard to political matters, it should first be pointed out that the number of democratic regimes is still quite low, while several territories are no longer under legitimate governmental control (there are cases of "delimited chaos" under the effective control of rebel/militarized groups). If, in general, democratization has made some important progress since the end of the 1980s, it is also true that several areas have also fallen into a situation of civil war. In fact, since decolonization, the Indian Ocean has been more a zone of war than a zone of peace as international conflicts and secessionist movements have been numerous and have involved every sub-region. If some progress has been made in the recognition of international borders, the pacification of border areas is still to be achieved in many places. In addition, there are direct implications from neighbouring states in terms of indirect support to a particular group, which often contributes to an exacerbation of national conflicts and creates regional instability.

Concerning ethno-cultural matters, these appear to be of primary importance, as the two main antagonisms of the Indianoceanic area are mostly driven by religious factors, although both religion and ethnicity are at the root of conflicts that arise at the national scale. Thus, the Israeli-Arab and Indian-Pakistan conflicts are intrinsically related to antagonisms between Jews and Muslims on the one hand, and between Muslims and Hindus, on the other hand. Elsewhere, the relations that are established between the majority groups and the minorities often represent a fundamental element of national political life. Finally, the presence of Indian, Muslim as well as Chinese communities all over the region represents an important cultural element that should contribute to an Indianoceanic identity. For us, even if ethno-cultural factors have fueled many conflicts until now, the great variety of peoples should be seen as one of the region's riches in an eventual peaceful and cooperative Indianoceanic world; if, of course, such a world is to develop.

In regard to economic matters, the Indianoceanic region clearly appears to be multi-polar and is dominated by its relations with the industrial Triad. The economic structure is essentially characterized by a very large disparity among national economies, the significance of natural resources and the weakness of intra-regional trade. The regional economic powers are India, Australia, Indonesia, Singapore, Thailand, Malaysia, Saudi Arabia, South Africa and Iran (Table 4.1). Together, these nine countries represent 72.9% of the Indianoceanic GNP and 78.0% of regional exports. The economic geography of the region is dominated by oil production in the Persian Gulf and the related maritime oil fluxes, the industrial production of the South-East Asian tigers that is intended to serve world markets, the Singapore commercial and financial crossroads, the mineral resources of Southern Africa, Australia and Indonesia (oil and gas included), the diversified economy of India and the size of its national market, as well as a great variety and the large volume of tropical products exported to the developed countries of the North.

Table 4.1. The Nine Economic Powers of the Indianoceanic Region.

	Percentage of the regional GNP	Percentage of the regional GNP expressed in PPP [a]	Percentage of regional exports	Indicator of economic power IEP [b]
India	16.41	30.50	5.10	17.34
Australia	15.68	7.11	8.89	10.56
Indonesia	8.73	10.87	7.78	9.13
Singapore	4.09	1.66	20.31	8.69
Thailand	6.78	6.66	8.19	7.21
Malaysia	3.93	3.89	11.79	6.54
Saudi Arabia	5.90	3.36	9.47	6.24
South Africa	5.39	5.25	4.55	5.06
Iran	5.98	6.13	1.94	4.68
The nine States	72.89	75.43	78.01	75.44

(a) In Purchasing Power Parity. (b) IEP represents the mean of all three percentages.
Data for 1998. Source: Bouchard, 2000, p. 71.

Finally, concerning military matters, the power of the armed forces from the Middle East and South and South-East Asia contrasts with the weakness of African forces. The race for modern armaments is a significant fact of recent decades in what we have called the "Indianoceanic arc of militarization", which stretches from Egypt to Indonesia and Australia. This strengthening of military power especially concerns naval and ballistic capabilities. At a time when the proliferation of weapons of mass destruction (WMD) considerably increases destructive capacity, the strategic area of the Indianoceanic states, which has been confined essentially to the land in the past, is now widening to the sea, air and even outer space. This makes regional conflicts even more dangerous and can complicate the manoeuvres of foreign militaries in the region. Without doubt, India is the largest Indianoceanic military power, but it does not possess the necessary means to control the whole of the Indian Ocean. Its armed forces must concentrate first on their mission of national territorial defence, especially in the context of the Indo-Pakistan rivalry, while paying more attention to Chinese developing interests in the Indian Ocean. Elsewhere, Afghanistan and Iraq are the object of American-led interventions and engagement in difficult state-rebuilding processes, while the United States has also been increasing pressure on Tehran about potential ballistic and WMD capabilities [4].

The Sub-Regional Systems

In the Indianoceanic Region, due to the generalized limited power of the states and thus the fact that their area of influence is usually constrained to their close neighbours, analysis of conflictual and cooperative dynamics must first be undertaken at the sub-regional scale. It is at this level that the main power relations are established, that the significant geopolitical equilibrium is formed as well as that the operative economic and political integration processes are organized. Thus, the second characteristic of the Indianoceanic region is its division into sub-regional systems. In this context, the Indian Ocean has often been qualified as a scattered geopolitical area or even has been denied the status of geopolitical area as Indianoceanic relations and interactions were so weak. Today, no contradiction is evident between the fact that there is a large-scale Indianoceanic system developing at the same time that sub-regional systems remain of principal significance. In the predictable future, the former is not to replace the latter; rather, it will rather complement them and enable a closer relationship among them.

On the Indian Ocean Rim, nine original and distinct sub-regional systems coexist, namely: Southern Africa, Eastern Africa, South-West Indian Ocean Islands, Horn of Africa, Persian Gulf, Central Asia, South Asia, South-East Asia and the very particular and remote area of the Austral Islands and EEZ (Figure 4.2). These sub-regional systems are bounded to larger systems such as those of the African Union, the Middle East, or the Asia-Pacific, and, of course, to the Indianoceanic system, as well as, at the highest scale, to the world system. Only six states are not included in one of these nine sub-regions, namely: Australia, East Timor, Egypt, Israel, Jordan and Yemen.

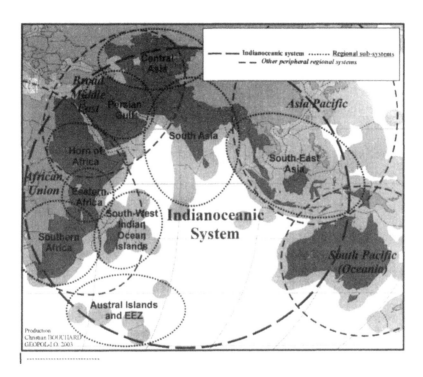

Figure 4.2. The Indianoceanic Regional Sub-systems
and Peripheral Regional Systems.

This sub-regional analysis shows that economic integration, which was very minimal at the beginning of the 1990s in most of the Indianoceanic Region, now represents a profound movement and is a rapidly developing process (despite several difficulties). This can be shown by the dynamism of the international organizations of regional vocation (Tables 4.2 and 4.3) that pursue the goal of establishing free-trade areas (SADC, ASEAN, EAC) or custom unions (GCC, COMESA; while the new SACU agreement entered into force in 2002). Other regional associations have developed significant sectoral cooperation and, even if advanced economic integration is not yet framed in a timetable, they have already formed a preferential trade area (SAARC, COI) or have progressed in this direction (IGAD) [5]. Economic integration is also promoted by some new initiatives which have now developed outside of the formal sub-regional framework already in place (for example, BIMST-EC, MGC and the Kunming Initiative [6]. As economic integration cannot be achieved without a minimum of political cooperation, regionalism contributes in several ways to the promotion of peace and security among associated states. However, on this matter, much remains to be done, as conflictual dynamics are still fundamental in most sub-regional systems.

With its three specific protocols [7] and its Regional Forum addressing peace, security and stability issues, ASEAN has developed an original framework which could serve as a model in the other sub-regions of the Indian Ocean. Elsewhere, even where conflicts are ongoing or tensions remain high, regional political cooperation is generally making some progress, as can be shown by the work of the SADC Organ on Politics, Defence and Security Cooperation, the Somalia and Sudan peace processes running under the auspices of the IGAD, as well as the GCC security agreement of 1994. On the other hand, multilateral peace and security cooperation is not yet on the agenda of either SAARC or IOR-ARC (India's position on this issue being largely responsible for that situation).

Table 4.2. Regional Associations in the Indianoceanic Region.

ASEAN	Association of Southeast Asian Nations
COI	Indian Ocean Commission
COMESA	Common Market for Eastern and Southern Africa
AEC	Commission for East African Co-operation
GCC	Gulf Cooperation Council or Cooperation Council for the Arab States of the Gulf
IGAD	The Intergovernmental Authority on Development
IOR-ARC	Indian Ocean Rim Association for Regional Cooperation
SAARC	South Asian Association for Regional Cooperation
SACU	Southern African Customs Union
SADC	Southern African Development Community

Emergent Indianoceanic Regionalism

The third characteristic of the Indian Oceanic Region is the emergence of a large-scale regional dynamic of relations and interactions. The concept of an Indian Ocean Rim (IOR) as a community of solidarity implies that a regional identity is developed and that regional cooperation is organized. It is in the conjunction of these two parallel processes that the IOR will find all of its substance. Presently, this large-scale regionalism does not represent a strong reality (political, economic or cultural) that dominates the Indianoceanic system, but it is definitively making progress, both from above (by the way of international relations that are realized and promoted by states) and from below (by the way of transnational relations that take the form of cultural, professional, humanitarian and economic networks). Thus, the Indian Ocean Rim is an emergent reality that, in all likelihood, constitutes a profound and probably irreversible movement. For this reason, Indianoceanic regionalism could well affirm itself as one significant element of the regional geopolitical framework in the twenty-first century. This evolution would be quite consistent with an actual generalized trend in international relations, as states can no longer leave aside their regional environment.

Established in 1997, the *Indian Ocean Rim Association for Regional Cooperation* (IOR-ARC) represents the institutional counterpart of Indianoceanic regionalism. It is essentially an economic process that has the goals of promoting inter-regional relations as well as of facilitating the integration of national economies into the world economy. At the moment, only those states which possess a port on the main ocean body (thus excluding the Persian Gulf and the Red Sea, but not the Malacca/Singapore Strait) have been invited to join the Association, a politic that therefore disables some important and influential coastal states such as Saudi Arabia from admission and marginalizes the land-locked states whose transit is oriented to the Indian Ocean. Whatever the economic objectives of IOR-ARC, Indianoceanic cooperation would have to be extended to peace and security, otherwise the ideal of the emancipation of the Indian Ocean Rim as an Indianoceanic community that effectively controls the region will never fully be realized. It is moreover on these latter issues that a regional shared interest in the Declaration of the Indian Ocean as a Zone of Peace (IOZP) has, for the first time, effectively and formally bound the Indianoceanic states together. Whatever the reasons for the failure of IOZP [8], academic works, diplomatic negotiations and political debates that have been achieved in relation to it have promoted the concept of an Indian Ocean community.

In addition to common concerns, the Indian Ocean Rim cannot be evoked without due consideration of the role and influence of the three Indianoceanic powers, namely, India, Australia and South Africa. These states are the only ones that possess all of the necessary attributes (political, economic, cultural and even military) to effectively exercise a certain leadership at the regional scale. All three play the card of South-South relations and try to gain something out of emergent regionalism: India pursing its old ambition of regional pre-eminence (with an eventual international recognition that could, for example, take the form of permanent membership of the UN Security Council), Australia looking mostly to seize good economic opportunities that are arising from developing economies (for example, the *Mozal Project* in Mozambique)

Table 4.3. Membership of Regional Associations in the Indianoceanic Region.

	COI	SACU[a]	SADC	EAC	COMESA[a]	IGAD	GCC	SAARC	ASEAN	IOR-ARC
Comoros	X				X					
Madagascar	X				X					X
Mauritius	X		X		X					X
Reunion, France for	X									
Seychelles	X		(w)		X					(w)
Botswana		X	X							
Lesotho		X	X							
(Namibia)		X	X		X					
South Africa		X	X							X
Swaziland		X	X		X					
(Angola)			X		X					
(D.R. of Congo)			X		X					
Malawi			X		X					
Mozambique			X							X
Tanzania			X	X						X
Zambia			X		X					
Zimbabwe			X		X					
Burundi					X					

Country	1	2	3	4	5	6
Djibouti				X	X	
Egypt				X	X	
Eritrea				X	X	
Ethiopia				X	X	
Kenya	X			X	X	X
Rwanda					X	
Sudan				X	X	
Uganda				X	X	X
Somalia				X	X	
Bahrain			X			
Kuwait			X			
Oman	X		X			
Qatar			X			
Saudi Arabia			X			
United Arab Emirats	X		X			
Bangladesh	X	X				
Bhutan		X				
India	X	X				
Maldives		X				
Nepal		X				
Pakistan		X				

	COI	SACU[a]	SADC	EAC	COMESA[a]	IGAD	GCC	SAARC	ASEAN	IOR-ARC
Sri Lanka								X		X
Burma (Myanmar)									X	
(Brunei)									X	
(Cambodia)									X	
Indonesia									X	X
(Laos)									X	
Malaysia									X	X
(Philippines)									X	
Singapore									X	X
Thailand									X	X
(Vietnam)									X	
Australia										X
Iran										X
Yemen										X

[a]SACU and COMESA are not formal regional associations but rather regional treaties respectively establishing a custom union and a common market/free trade area. In brackets: States that are not considered as belonging to the Indianoceanic Region. Dialogue partners and observers are not taken into account in this table. In addition, the Cross-Border Initiative (CBI), which includes 14 African States, seeks a market driven concept of integration in Eastern and Southern Africa and the Indian Ocean Islands countries. (w) Facing a financial crisis, Seychelles withdrawn from SADC and IOR in 2003.

and emergent markets (for example, the United Arab Emirates), and South Africa extending its interests and influence far above its traditional privileged area covering Southern Africa and the South-West Indian Ocean Islands. However, these three large regional powers are not the only Indianoceanic states able to exercise some influence on the region, especially on economic matters (Table 4.1), nor are they the only ones exercising strong regional diplomatic leadership (for example, Mauritius [9]).

Outside of the formal actions of states, associations of every kind increasingly network the region. Some like the *Indian Ocean Rim Business Forum* (OIRBF), the *Indian Ocean Rim Academic Group* (IORAG) and the *Indian Ocean Tourism Organization* (IOTO, as observer) are linked to IOR-ARC, and thus should play a role in orienting the future development of Indianoceanic regionalism. However, the other associations are as important as they link more and more individuals from all around the region, and therefore contribute to the promotion of a wide regional identity. In this particular context, three different solidarities can be exploited as the root of this regional identity:

- First, the promotion of South-South relations and common interests which are developed in reaction to both the colonial past and the actual economic domination of the North, and thus implying an emancipation from the influence of the large foreign powers;
- Secondly, the Indian diaspora, which represents a significant Indianoceanic actor that has already developed commercial, financial and cultural networks covering the whole region, and acts at the same time as an effective means of diffusion for India's influence;
- Thirdly, Islam, which is the most common religion of the region, and especially in a crescent that goes from Comoros Islands to Malaysia and Indonesia and covers the Middle East, which is a cultural factor that could eventually run counter to the Indian goal of regional pre-eminence and, eventually, in promoting emancipation from the Western world and the large developed forei?n powers.

If these three regional solidarities are far more than only potential, then they are also limited by the fact that none of them is neither specific to the region nor of primary significance in the whole region. The admission of France for Reunion and Mayotte will probably not change the general situation in regard to these solidarities, but, if the Red Sea and Persian Gulf coastal states (especially Egypt and Saudi Arabia) were to be more closely bounded to Indianoceanic regionalism and gain IOR-ARC membership together with Pakistan, then Indian and Muslim influences could find fertile ground for competition, if not confrontation, in addition to the desired cooperation. At the same time, leaving those states out of the regionalization process will mean that some large regional players will be left aside, and therefore the strength of the regional association will be diminished. Thus, considering the main solidarities on which Indianoceanic regionalism can evolve, ideological and political issues are very significant, even if economy is the official objective of IOR-ARC.

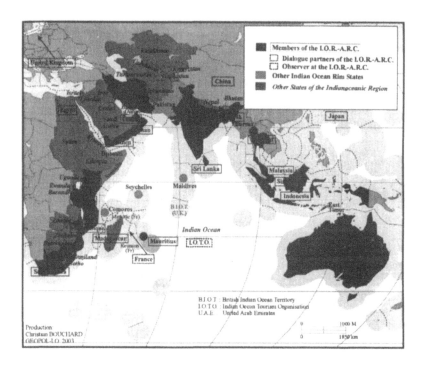

Figure 4.3. Indian Ocean Rim: Broad and Narrow Memberships.

The Subordination to Large Foreign Powers

Subordination to the large foreign powers represents the fourth characteristic of the Indianoceanic region. This is very meaningful because of the conjunction of two factors. First, the domination of the global system on the Indianoceanic system – inescapable because of the relative weakness of Indianoceanic states compared to the large world powers - means that the regional economic and even geopolitical contexts are strongly dependent upon the global system on which they often exercise a minimal influence. Second, the extrovert character of the region, which is very much an outcome of the colonial heritage, means that the Indianoceanic states maintain more important relations with foreign states than among them. In these conditions, the influence of large foreign powers is predominant and multifaceted.

Table 4.4. The Large Foreign Powers' Influence in the Indianoceanic region.

Field	Predominant	Important	Weak	Very weak or non-existent
Military (hard power)	United States	United Kingdom France	China Russia	Japan Germany
Economy (economic power)	United States Japan European Union*	United Kingdom Germany France China	Russia	
Culture (soft power)	United States United Kingdom		France China	Russia Germany Japan
Diplomacy (political power)	United States	United Kingdom Russia France China	Japan	Germany

*The large European powers have maintained specific military and cultural influences while their economic influence is clearly more collect-ive as members of the European Union.

On the one hand, this influence is predominant in the context where the actual regional equilibrium is largely related to the large powers' actions and interests and to which it is globally quite favourable. This is a fundamental point, since, for the industrial Triad, the region is both a vast resources reservoir to exploit and a large market to capture, two essential facts for the good functioning and wealth of the dominant developed industrial economies. On the other hand, this influence is multifaceted as the general influence of the large foreign powers is the result of various factors, ranging from economic to cultural, military and diplomatic. The specific influence of each large foreign power on the Indianoceanic system depends especially on its interests in the region, the means it disposes to promote and defend them, as well as the will it demonstrates to secure them.

As the main attributes of state power can be evoked as hard power (military capabilities), economic power (trade, finance, foreign direct investment, international aid), soft power (cultural production, control of the means of diffusion) and political power (diplomatic weight, capabilities to extend traditional diplomacy by economic or even military actions), it is therefore possible to analyze the specific influence of each great foreign power through its different power manifestations. As there is no absolute measure of these manifestations, the results are best expressed in terms of relativity and in the form of a comparative table (Table 4.4). It is also important to note that the proposed classification of foreign influence is representative of the actual situation, and thus, has greatly evolved since the end of the Cold War and should continue to evolve in the future.

The Preponderance of the Indian Ocean

The Indian Ocean itself is far more than just the geometric centre of the region. If we cannot assert that it is its geographical centre, as the regional poles are all located on its shores, it is clearly not a simple periphery. In fact, the Indian Ocean is the core element of the region as the sea is both the regional hyphen and therefore its fundamental unifying factor, and its major door to the outside, and, therefore, its main interface with the rest of the world.

If the Indian Ocean "under-maritimization" is a fact [10], this does not mean that maritime affairs are negligible for Indianoceanic states. On the contrary, this situation appears to be related to the generalized lower level of economic development in the region, to its secondary role in the world economy dominated by the large economic powers of the industrial Triad, as well as to the weakness of intra-regional exchanges; three features that are largely bounded to the colonial heritage. It is also important to nuance this global under-maritimization as, at national and local scales, the sea often constitutes an element of great significance. For example, we can evoke the oil states of the Persian Gulf and the regional commercial powers (Singapore, Malaysia, Thailand and Indonesia), the former for their exploitation of offshore resources and the latter for the role of maritime fluxes in their economy, as well as the thousands of coastal communities scattered all around the area which rely fully or partly on the sea for their living.

The major maritime issues, factors of primary significance in the Indianoceanic geopolitical framework, are:

- The territorialiszation of the sea, which is largely incomplete as only 28 maritime borders out of 65 potential borders have been the object of a formal delimitation (often only unifunctional), a situation that is complicated by the remaining insular territorial disputes [11]: Mayotte (between France and Comoros), Chagos (between United Kingdom and Mauritius), Abu Musa (between Iran and U.A.E.), Tromelin, Bassas da India, Europa, Gloriosos and Juan de Nova (between France and Madagascar), Nok, Lam and Khan (between Burma and Thailand), Batu Putish and Pisang (between Singapore and Malaysia);

- The fluidity and the security of maritime transport, especially in the strategic passages where the main fluxes concentrate (Suez Canal, Bab el Mandeb, Ormuz, Malacca/Singapore, Sunda and Lombok straits), and the serious problems of piracy (especially in South East Asian waters and off the coasts of the Indian sub-continent and the Horn of Africa), smuggling (for example, general products, people, drugs and arms), and, even now of maritime terrorism (especially in the north-west Indian Ocean);

- The exploitation of marine resources (biologic, mineral and energetic), which are very unequally distributed and are largely controlled by the coastal states as these enjoy exclusive rights on the resources of their EEZ and continental shelf, but extend to general concerns such as in the case of the tuna fisheries regulated by the *Indian Ocean Tuna Commission* (IOTC) or the illegal fishing activities recorded in several areas (for example, in the EEZ of the Austral Islands);

- The military use of the sea, as nothing in the international law formally bans foreign naval manoeuvres outside the territorial sea of any coastal state and warships of all states enjoy the right of transit passage in the international straits (UNCLOS, Art. 38) [12], a situation that allows the large naval powers to position, in pursuit of their geopolitical goals and strategic needs, warships all over the Indian Ocean, and gives them great operational liberty;

- The transit from and to the sea of the land-locked states, which necessitates the collaboration of the transit states, adequate communication networks and the transit security, without which the enclave situation is a strong developmental constraint for any land-locked states;

- The marine environmental degradation, that threatens water quality, marine biological resources and coral reefs (for example, in the Red Sea), and therefore can negatively affect fishing (for example, Persian Gulf) and tourism activities, as well as sea level rise that would severely impact upon low-lying islands and lands (for example, the Maldives, Mouths of the Ganges).

Conclusion

With the decolonization that followed the Second World War, the Indian Ocean Basin entered into a new geopolitical era. One after the other, the populations of the region regained political control over their territories in a quest for national independence that led to the creation of the modern Indianoceanic Asian, African and Islands states. However, as meaningful as it was, this was only the first step on the way to emancipation from the domination of large foreign powers. In the 1990s, being now free of the constraints of the Cold War, a second very important step was to be achieved as economic integration and political cooperation could become more dynamic and efficient among neighbours of the same geographical area and while an old wish of Indian Ocean Rim cooperation could finally be realized. If the former is in several ways related to the globalization that forces neighbours to face together common problems and develop their complementarities, the latter is the concrete manifestation of an emergent large-scale regionalism that is building upon the concepts of Indian Ocean community, identity and solidarity.

Thus, the populations of the region are now in the process of regaining some leadership and control over their region at both the neighbourhood and the Indianoceanic scales. In this new context, emancipation continues to progress as some Indianoceanic influential states now have a real role to play and exercise leadership at both the sub-regional and the regional scales. Nevertheless, in all likelihood, the Indianoceanic Region and its sub-regions will continue for a long time to be greatly influenced by foreign large powers, and especially those of the industrial Triad. The Indianoceanic order that succeeded the colonial order is strongly influenced by the new place that the Indianoceanic states and populations take in the geopolitical framework of the region. If it is not possible to know how far this emancipation trend will go, it is quite conceivable that what has been gained to date will remain as a fundamental element of Indian Ocean geopolitics in the 21st century.

With this emancipation process as a backdrop, our analysis has identified five main characteristics for the new Indianoceanic order, which are:

1) The great political, cultural and economic heterogeneity of the region;

2) A fragmentation into well-affirmed sub-regional systems, where economic integration and political cooperation are first organized and the geopolitical equilibrium is constructed;

3) An emergent Indianoceanic regionalism, which is original in nature and is now formalized by the *Indian Ocean Rim Association for Regional Cooperation* (IOR-ARC);

4) A subordination to large foreign powers, especially to those of the industrial Triad (United States, European Union and Japan) which have a tremendous influence on the region even if they do not exercise complete control of it; and

5) The very importance of the Indian Ocean itself, as it represents both the hyphen of the region and its main door to the rest of the World.

Finally, this Chapter has demonstrated that Indian Ocean geopolitics is far more than specific oil issues and the military uses of the sea by the large foreign powers. It is multifaceted in nature and influenced by factors that operate at all scales, from local to global. It is especially rich in conflictual and cooperative dynamics at the sub-regional level. As one of its new features, emergent Indianoceanic regionalism raises many new issues for the 21st century. However, if there is a clear quest for more cultural, economic, political, and even strategic emancipation from the large foreign powers, then this does not necessarily mean that broad and dynamic regional cooperation will be successfully developed.

Notes

[1] According to the limits addressed by the International Hydrographic Organization, the Indian Ocean formally includes the Red Sea, the Persian Gulf, the Gulf of Aden, the Gulf of Oman, the Mozambique Channel, the Arabian Sea, the Laccadive Sea, the Gulf of Mannar, the Palk Bay, the Bay of Bengal, the Andaman Sea, the Strait of Malacca and the Great Australian Bight. On the other hand, it excludes the Java Sea, the Banda Sea, the Arafura Sea and the Timor Sea as well as the Indonesian straits, the Torres Strait and the Bass Strait.

[2] *Report of the Ad Hoc Committee on the Indian Ocean*, GAOR, 34th sess., Suppl. no. 45 (A/34/45). Even if it proposed a geographical delimitation for the Indian Ocean as a zone of peace, this report states that "the final limits of the Indian Ocean as a zone of peace have yet to be agreed upon", an agreement that has not yet materialized.

[3] Some prefer to call it the Arabo-Persian Gulf as it is not uniquely bordered either by Arabs or Persians; others call it theArabian Gulf.

[4] If Iraq and Iran have been pinpointed as the most dangerous threat to regional peace and stability as they possess both ballistic missiles and weapons of mass destruction capabilities, we should not forget that India,

Israel and Pakistan are nuclear powers and that several other Indianoceanic states possess ballistic missiles.

[5] Regional Associations on the Internet:

ASEAN: www.aseansec.org

COI: www.coi-info.org

COMESA: www.comesa.int

EAC: www.tanzania-online.gov.uk/eacommunity.html

GCC: www.gcc-sg.org

IGAD: www.igad.org

SAARC: www.saarc-sec.org

SACU: www.dfa.gov.za/for-relations/multilateral/sacu.htm

SADC: www.sadc.int

[6] BIMST-EC: Bangladesh, India, Myanmar, Sri Lanka, Thailand – Economic Cooperation; MGC: Mekong Ganga Cooperation (Cambodia, India, Laos, Burma, Thailand and Vietnam). The Kunming Initiative interlocks India's northeast region with Bangladesh, Burma and southwest China.

[7] The Treaty of Amity and Cooperation (TAC) in Southeast Asia, the Zone of Peace, Freedom and Neutrality (ZOPFAN) Declaration, and the Treaty on the Southeast Asia Nuclear Weapon-Free Zone (SEANWFZ).

[8] The concept of the Indian Ocean as a Zone of Peace (IOZP) has not been officially abandoned by the United Nations. The Ad Hoc Committee on the Indian Ocean is still alive and its Chair continues informal consultations on the matter. Even if the IOZP remains on the official agenda of the General Assembly, there is no sign that it will be possible to go ahead with the concept definition and implementation in a foreseeable future (Bouchard, 2003),.

[9] Mauritius has played an essential role in the promotion of a formal Indianoceanic regional association. After several consultations with other states in the region, and especially the three Indianoceanic powers, Mauritius launched the Indian Ocean Rim Initiative (IORI) in 1995, which eventually led to the establishment of the IOR-ARC in 1997. In 2000, the IOR-ARC Coordinating Secretariat opened in Mauritius. Other states do have influence, for example, Sri Lanka in the IOZP process, and Malaysia for the promotion of South-South cooperation, Asia-African relationship and Islamic solidarity.

[10] In an extensive work on sea and the geostrategy of sates, Vigarié (1995) identifies the under-maritimization and the richness of transit as the two main characteristics of the Indian Ocean Region. For this author, the Indian Ocean is the ocean of under-development which is not only manifest on its littoral, but also at sea as a result of a relatively weak participation to maritime transport and the small and insufficient exploitation of marine resources. Globally, Indian Ocean under-maritimization is expressed in the small intra-regional level of exchange, the small size of the national fleet tonnage in relation to the number of inhabitants, and we will add the limited naval and coastguard capabilities of several states.

[11] In a recent Judgment (17 December 2002), the International Court of Justice stated that the sovereignty over Pulau Ligitan and Pulau Sipadan belongs to Malaysia.

[12] The United Nations Convention on the Law of the Sea (UNCLOS) was adopted in 1982 and entered into force in 1994.

References

Bouchard, Christian (2000), *L'espace indianocéanique, un système géopolitique en recomposition.* Quebec, Canada: Laval University, Department of Geography, Ph.D. thesis.

Bouchard, Christian (2003), "The case for security and strategic matters in the Indian Ocean cooperation", Hyderabad: Osmania University, Centre for Indian Ocean Studies, *Indian Ocean Digest*, vol. 17, no. 2, July-December 2002, pp. 13-21.

CIA (2003). *The World Factbook 2003.* http://www.cia.gov/cia/publications/factbook/index.html.

Forbes, Vivian L. (1995), *The Maritime Boundaries of the Indian Ocean* (Singapore: Singapore University Press).

Lauseig, Jérôme (2000), "L'Indian Ocean Rim à l'épreuve de la éopolitique, contribution à l'étude des formes nouvelles du régionalisme" *in* Institut Austral de Démographie (I.A.D.). *Actes du séminaire du 26 novembre 1999* (Reunion: Observatoire du Développement de la Réunion (O.D.R.), pages 51-69.

Maurice, Pierre and Gohin, Olivier (ed.) (1994), *International relations in the Indian Ocean* (Reunion: Université de la Réunion, Centre d'Études et de Recherches en Relations Internationales et Géopolitique de l'Océan Indien (CERIGOI)), pages 219-231.

McPherson, Kenneth (1998), *The Indian Ocean, a History of People and the Sea* (Delhi: Oxford University Press, Oxford India Paperbacks).

PRB (2003), *World Population Data Sheet* (Washington: Population Reference Bureau).

Rao, P. V., ed. (2003), *India and the Indian Ocean in the Twilight of the Millenium: Essays in Honour of Professor Satish Chandra* (Delhi: South Asian Publishers).

United Nations (1979), *Report of the Meeting of the Littoral and Hinterland States of the Indian Ocean* (New York: General Assembly, 34th session, supplement No. 45 (A/34/45)).

UNDP (2003). *Human Development Report 2003, Millennium Development Goals: A Compact Among Nations to End Human Poverty* (New York: Oxford University Press).

Vigarié, André (1995), *La mer et la géostratégie des nations* (Paris: Economica and Institut de stratégie comparée (ISC), Coll. Bibliothèque stratégique).

Wriggins, W. Howard, ed. (1992), *Dynamics of Regional Politics: Four Systems on the Indian Ocean Rim* (New York: Columbia University Press).

Appendix 4.1. Basic Data for States and Territories of the Indian Ocean Region.

	Total area (sq.km)	Population mid-2003 (millions)	Absolute density (inh./sq. km)	H.D.I. 2001	G.D.P. Est. 2001 (billions PPP US$)
Afghanistan	647 500	28.7	44	–	–
Australia	7 686 850	19.9	3	0.939	491.8
--- Ashmore & Cartier	5	–	–	–	–
--- Christmas I.	135	0.002	15	–	-
--- Cocos Is (or Keeling)	14	0.001	71	–	–
--- Heard & McDonald	412	–	–	–	–
Bahrain	665	0.7	1 053	0.839	10.5
Bangladesh	144 000	146.7	1 019	0.502	214.1
Bhutan	47 000	0.9	19	0.511	3.8
Botswana	600 370	1.6	3	0.614	13.3
Burma	678 500	49.5	73	0.549	49.5
Burundi	27 830	6.1	219	0.337	4.8
Comoros	2 170	0.6	276	0.528	1.1
Djibouti	23 000	0.7	30	0.462	1.5
East Timor	15 007	0.8	53	–	–
Egypt	1 001 450	72.1	72	0.648	229.4
Eritrea	121 320	4.4	36	0.446	4.3
Ethiopia	1 127 127	70.7	63	0.359	53.3
France	–	–	–	–	–
--- Crozet	325	–	–	–	–
--- Indian Ocean Sparse Islands	30	–	–	–	–
--- Kerguelen	7 215	–	–	–	–
--- Mayotte	374	0.2	535	–	–
--- Reunion I.	2 517	0.8	318	–	–
--- Saint-Paul & Amsterdam	61	–	–	–	–
India	3 287 590	1,068.6	325	0.590	2 930.0
--- Andaman & Nicobar	8 249	0.3	36	–	–
--- Lakshadweep	32	0.05	1 563	–	–
Indonesia	1 919 440	220.5	115	0.682	615.2
Iran	1 648 000	66.6	40	0.719	387.2
Iraq	437 072	24.2	55	–	–
Israel	20 770	6.7	323	0.905	125.9
Jordan	92 300	5.5	60	0.743	19.5
Kazakhstan	2 717 300	14.8	5	0.765	96.8
Kenya	582 650	31.6	54	0.489	30.1
Kuwait	17 820	2.4	135	0.820	38.2
Kyrgyzstan	198 500	5.0	25	0.727	13.6
Lesotho	30 355	1.8	59	0.510	5.0

	Total area (sq.km)	Population mid-2003 (millions)	Absolute density (inh./sq.km)	H.D.I. 2001	G.D.P. Est. 2001 (billions PPP US$)
Madagascar	587 040	17.0	29	0.468	13.3
Malawi	118 480	11.7	99	0.387	6.0
Malaysia	329 750	25.1	76	0.790	208.3
Maldives	300	0.3	1 000	0.751	1.4
Mauritius	2 040	1.2	588	0.779	11.8
Mozambique	801 590	17.5	22	0.356	20.6
Nepal	140 800	25.2	179	0.499	30.9
Oman	212 460	2.6	12	0.755	29.0
Pakistan	803 940	149.1	185	0.499	266.7
Qatar	11 437	0.6	52	0.826	11.9
Rwanda	26 338	8.3	315	0.422	10.9
Saudi Arabia	1 960 582	24.1	12	0.769	285.3
Seychelles	455	0.1	220	0.840	1.3
Singapore	693	4.2	6 061	0.884	93.7
Somalia	637 657	8.0	13	–	–
South Africa	1 219 912	44.0	36	0.684	488.2
--- Prince Edward Is.	334	–	–	–	–
Sri Lanka	65 610	19.3	294	0.730	59.6
Sudan	2 505 810	38.1	15	0.503	62.3
Swaziland	17 363	1.2	69	0.547	4.6
Tajikistan	143 100	6.6	46	0.677	7.3
Tanzania	945 087	35.4	37	0.400	18.0
Thailand	514 000	63.1	123	0.768	391.7
Turkmenistan	488 100	5.7	12	0.748	23.5
Uganda	236 040	25.3	107	0.489	33.9
United Arab Emirates	82 880	3.9	47	0.816	59.5
United Kingdom	–	–	–	–	–
--- British Indian Ocean Territory	60	–	–	–	–
Uzbekistan	447 400	25.7	57	0.729	61.6
Yemen	527 970	19.4	37	0.470	14.3
--- Socotra	3 626	0.05	–	–	–
Zambia	752 614	10.9	14	0.386	8.0
Zimbabwe	390 580	12.6	32	0.496	29.3

Major sources: PRB, *2003 World Population Data Sheet* (population); CIA, *The World Factbook 2003* (total area); UNDP, *Human Develop-ment Report 2003* (HDI and GDP); GDP estimated from population and GDP per inhabitant for Bhutan, Burma, Maldives, Qatar, Seychelles and United Arab Emirates.

REGIONALISM

CHAPTER 5

"There Once was an Ugly Duckling…," or, The Sad History of the Good Ship IOR-ARC

Kenneth I McPherson

Introduction

In 1997, 14 countries on the shores of the Indian Ocean signed a Charter that established the Indian Ocean Rim Association for Regional Cooperation (IOR-ARC).

The signing of the Charter came after three years of often tense, and sometimes acrimonious, negotiations and was meant to herald an era of cooperation across a region that stretched from East Africa to Southeast Asia, and which was home to a third of the world's population.

The extent of this anticipated cooperation was vaguely outlined in the Charter, but was posited principally on a desire to promote greater economic interaction and human resource development across the region as a means of improving the livelihood of the inhabitants of what is paradoxically both the world's poorest and richest region. In terms of per capita income, the Indian Ocean region lags behind the rest of the world, yet, in terms of natural resources, it is very much the quarry of the so-called "developed world", providing, as it does, many of the resources that underpin the economic life of the industrialized states of the northern hemisphere.

It was anticipated that the new Association would be a tripartite organization comprising officials, business representatives and academics, who would collectively work to address issues of common concern as outlined in the Charter. The day-to-day running of the Association was in the hands of a Secretariat located in Mauritius.

The model for this type of approach was the Asia Pacific Economic Cooperation (APEC) grouping, with its successful track record – at that time – of tripartite activity and interaction. There was also a second track of activity comprised of separate academic and business groupings made up of participants from the 14 IOR-ARC member states and participants from states not yet members of the Association. This more inclusive – in terms of membership – second track grouping was recognized by IOR-ARC as operating in support of the Association's objectives.

The Creation of IOR-ARC: Internal Vision and Reality

In the years immediately following the establishment of IOR-ARC there was a flurry of meetings in Mozambique and Oman, but by the turn of the century – barely three years after the establishment of the association – it had changed dramatically from the organization envisioned in 1997. The second track groupings had vanished and the business and academic components of the tripartite governing mechanism were, in effect, moribund. The association lived on, but it had changed into an organization dominated by trade and foreign ministry officials with a very sharp focus on exploring issues relating to trade and investment facilitation across the Indian Ocean region.

For some, given the eloquent speeches made between 1994 and 1997, what existed by 2000 bore little resemblance to what had been anticipated in the first rush of enthusiasm for regional cooperation. As a participant in the numerous meetings of that period, I must admit that I look on the present Association with a sense of disappointment, and, in my worst moments, see it as having been high-jacked by faceless 'economically rational' trade bureaucrats determined to dehumanize rather than humanize any process of regional cooperation in their eagerness to avoid anything that might cut across other agendas. This is obviously an emotional and perhaps an unfair conclusion: perhaps the IOR-ARC of today is as much regional cooperation as the region can comfortably sustain in the present circumstances. Certainly – I have to grudgingly admit – the bureaucrats are not the villains, there are in reality no villains in this story. The real problem facing Indian Ocean regional cooperation has always been the absence of any clear vision of what such cooperation could and should entail and the relatively weak commitment to the concept of regional cooperation by the member states of IOR-ARC.

I can feel the hackles rising in certain quarters as I make such a bold statement. Given the voluminous speeches concerning regional cooperation and the action-packed meetings that ushered in IOR-ARC, how can I make such claims?

What were the visions? They ranged from the vision of Australia's Gareth Evans of an inclusive regional organization on the lines of APEC and ASEAN that would explore and promote a range of issues with particular emphasis on economic cooperation and "confidence building", to the more sober visions of Indian and South African spokespeople whose focus was almost entirely upon economic cooperation among a select group of Indian Ocean states. From the beginning, there was remarkably little debate about mechanisms of government for any proposed organization and the consensus was for a de facto adoption of the tripartite governance model of APEC.

IOR-ARC Membership

There were from the beginning obvious differences relating to membership and to the objectives of any proposed Indian Ocean regional organization. These were the core issues that consumed time and energy in the debates from 1994 until IOR-ARC was formed in 1997. Here it needs to be noted that neither the issue of governance nor the relevance of the APEC model were debated in any depth.

The discussions between 1994 and 1997 focussed on three main issues: (1) the question of membership, that is, the well worn "inclusive" versus "exclusive" debate; (2) the debate between proponents of an exclusive "trade bloc" and the proponents of "open regionalism"; and (3) the scope of issues to be addressed by the proposed organization.

1)   This was a sensitive topic with Australia leading the charge for inclusive membership and India arguing for keeping it small. India argued that a smaller group of committed states would be more effective than a larger amorphous group. It argued that this had been the path chosen by APEC (check mate Australia!) but there also appeared to be an element of wishing to keep Pakistan out of the organization. The eventual compromise was for a gradually expanding membership with Pakistan able to apply for membership once it had regularized its international trading regime particularly with respect to extending the "most-favoured-nation" treatment to India.

2)   On the question of trade, given the commitment of the leading economies in the region to the WTO, it was inevitable that "open regionalism" would have to be adopted.

3)   Some issues were seen by some Association members as not be included on the IOR-ARC agenda. For example, strong arguments were put that security issues and bilateral disputes were "no go"

areas for the organization. It was finally agreed that discussion of domestic and security issues was not within the scope of the proposed organization and a strong emphasis was placed on the role of the organization in facilitating trade and investment among member states and promoting human resource development.

IOR-ARC Structure and Function

By 1997, broad agreement had been reached on the scope of the organization's activities, a system of governance was agreed upon, and IOR-ARC was launched.

But what was this creature that had been launched? Its objectives and raison d'être were vague, and its system of governance was not clearly spelled out.

Let me take the problem of governance. A weak Secretariat was established whose role was far from clear and whose relationship with the three governing groups was not defined. Furthermore, there was no agreement on how the tripartite governing system was to work. There was no mechanism in place for the establishment and maintenance of either the academic or business groups and the only self-perpetuating part of the tripartite governing system was the official bloc.

For the first year or so, the momentum of IOR-ARC was sustained by the initial flush of enthusiasm, but, by 1998, the wheels began to fall off the cart. The problems for IOR-ARC were twofold – internal and external.

Internally, the immediate problem was the relationship among the groups comprising the tripartite leadership. Put simply, neither the academic nor business groups flourished.

Of the two, the business group was undoubtedly the most active with strong support from the Indian and Omani business communities in particular. However, the proactive role of Indian business organizations, such as FICC and the CII, was the exception rather than the rule and business participation from Australia, Singapore and South Africa remained disappointing with government in these states providing the main input. Academic representation was even more abysmal. A core of states provided some representation but most member states simply did not bother, or were represented in this group by government officials. The end result was that within a couple of years of the establishment of IOR-ARC, governance of the organization had, in effect, become the prerogative of the official group, with the academic and business groups exercising minimal influence.

Certainly some member states of IOR-ARC attempted to honour the terms of the Charter with respect to the tripartite nature of the organization, but the majority of member states appear to have been content to leave their representation solely in the hands of officials.

IOR-ARC Governance

In part, this appears to have been caused by the failure of many member states to understand the tripartite system of governance, but it was also caused by the initial uncritical acceptance of the APEC model. What was not remembered was that the APEC structure was evolutionary in origin. Long before the establishment of APEC, business and academic groups in the Asia-Pacific region had begun to form regional groupings to discuss common interests. These groups, in turn, influenced government policy and were the foundations upon which officials launched APEC. The situation in the Indian Ocean region was quite different. There was no background of regional business cooperation, and there was only marginally more academic cooperation. Neither group had produced a single regional pressure or interest group. When officials in the region became enamoured of the idea of regional cooperation they created an institution from above – an institution which lacked any roots in the region. This meant that, from its inception, IOR-ARC had a very limited constituency and there was a pressing need to foster an academic and business interest.

In the period from 1994 to 1997, and for a few years after, some IOR-ARC member states did attempt to foster a proactive business and academic input but most member states honoured the concept more in the breach than in the observance, and, by 2000, the academic and business components of the tripartite system were all but moribund, apart from some officially-sponsored and irregular input. Whilst official representation had a continuity and substance, the same could not be said of the academic and business groups that were crudely cobbled together to meet the immediate requirements of particular IOR-ARC meetings.

If the world had remained static, or at least had not thrown up any unexpected events, then IOR-ARC may have had time to face and deal with its internal problems, but, unfortunately, this was not to be the case.

IOR-ARC: External Linkages and Impacts

In the world outside IOR-ARC several events and developments undermined interest in the organization.

The first of these was the East Asian economic crisis of 1997 which distracted IOR-ARC's Southeast Asian members to the extent that it even threatened ASEAN and impacted negatively on APEC.

This event in itself may not have impacted all that severely upon IOR-ARC if it had not occurred at a time when there was a growing trend within some sections of the international order away from multilateralism and towards bilateralism.

At this point, might I suggest that the late 1990s will be marked by commentators in the future as a period when the growing hegemony of the USA ushered in some fundamental changes in international relations. The most notable of these was a discounting of multilateralism and multilateral organizations in favour of bilateralism. Such a development received further encouragement as the world economy hovered on the brink of recession and individual states moved to secure their economies. In Southeast Asia, individual states in part turned from ASEAN in search of bilateral free trade agreements elsewhere, and, across the Indian Ocean region, there was a general rush to establish free trade agreements with major extra-regional economies...not let it be noted with one another, except in the instance of Australia and Singapore.

None of this augured well for IOR-ARC, and, by the end of the 1990s, there was a clear decline in interest in IOR-ARC on the part of its three major protagonists: Australia, India and South Africa.

Whilst both India and Australia have maintained an interest in, and support for, IOR-ARC, this interest and support has undeniably declined. Both countries initially committed resources to support the tripartite structure and actively campaigned to breathe life into the concept of regional cooperation, but the extent of that support waned rapidly after 1997. Perhaps the single most important reason for this has been the changing global economic and security environment with the emergence of a single hegemonic Superpower and the decline in influence of multilateral organizations.

Both Australia and India appear to have moved away from a commitment to multilateralism and have developed a keener interest in reinforcing their bilateral relationship with the USA: a process reinforced by the events of September 11, 2001, and the subsequent invasion of Afghanistan. As they have moved to develop this bilateral relationship, there has been a concomitant decline in their interest in IOR-ARC. The same has been true for all of the member states of IOR-ARC and although the number of member states has grown quickly since 1997, there has been a rapid decline in real commitment to making IOR-ARC function effectively.

In some ways, the case of South Africa may illustrate the real problems facing the majority of Indian Ocean states with respect to supporting IOR-ARC. Unlike the Asia-Pacific region, the Indian Ocean region does not include a large number of major world economies. The degree of resource commitment to any regional organization is constrained by the relative poverty of the Indian Ocean region and the pressing urgency of its economic and development problems.

Initially, South Africa was a keen supporter of the concept of regional cooperation and, indeed, pushed the concept ahead of Australia and India. In part, while the South African initiative was also driven by a general desire "to come in from the cold" after the collapse of the apartheid regime, in part it was driven by a hard headed assessment that South Africa needed to explore new markets and sources of capital investment, particularly in Southeast Asia and Australia. At the earliest meetings held to explore the formation of a regional organization there was strong government and business support from South Africa, but, in subsequent years, that interest steadily waned. In part, this was due to a hard-headed reassessment by government and business groups in South Africa concerning limited human resources and national priorities. The result was that the Indian Ocean region slipped down the list of national priorities way below negotiations with the EU and the USA. By 1997, support may have lingered for the creation of IOR-ARC, but it was far from being the main game in town and had been overtaken by more pressing issues and more promising bilateral relationships.

Conclusion

This waning of enthusiasm for IOR-ARC has been reflected around the region and one is now forced to ask if IOR-ARC has anything to offer other than as an official annual forum at which regional trade ministers meet? I would argue that it does, but only if it reinvents itself by broadening its basis for cooperation. In attempting to make IOR-ARC a mean, sleek and efficient machine, it was reduced to almost nothing. Yes, trade and investment are vital issues, but they are not necessarily the issues over which one can readily reach agreement, establish programmes for cooperative action and encourage habits of dialogue. Nor are they, let us be honest, issues which will secure the support of the public. The Indian Ocean region is beset by many problems that do stir the public mind and imagination, ranging from the HIV/AIDS pandemic to issues of comprehensive security. There is plenty of meat on the bone of regional dialogue and cooperation but participants to date have been offered a starvation diet with the result that they have lost interest in IOR-ARC and do not see it as an effective mechanism for promoting regional cooperation. Until regional governments are prepared to once more address the issue of IOR-ARC's agenda and commit resources to ensure a positive commitment on the part of regional business and academic groups, the organization will remain ineffective. There is an elephant's graveyard of dead and dying multilateral organizations somewhere out there full of weakly quacking ugly ducklings barely staying afloat. If the good ship IOR-ARC is to survive, it needs to be reshaped and given new impetus by governments currently prepared to give it little more than lip service.

CHAPTER 6

Sub-Regional Cooperation in South Asia: The Kunming Initiative

P.V. Rao

Regional Cooperation in South Asia has gathered momentum in recent years. India's market reforms and the compulsions of globalization spurred the South Asian countries to strengthen the SAARC system. A network of economic arrangements within and outside the SAARC framework has been designed to meet the demands of a growth economy and challenges of megablocs. The resolve to lift regional cooperation out of the stoicism of the past and the region's political climate is very evident, more often among the smaller states. Alternative regional development strategies are worked out instead of relying only on the collective sense of the seven-member SAARC, which is conditioned by the political vagaries of the region. To emphasize, states other than India and Pakistan are pushing for serious measures. SAPTA, (South Asian Preferential Trade Area) for example, is a Sri Lanka-Bangladesh initiative and SAGQ (South Asian Growth Quadrangle) is a Nepal-Bangladesh proposal. South Asia is thus pursuing three simultaneously overlapping strategies for deepening the regional cooperation process. These are: (1) Bilateral (for example, the Indo-Sri Lanka Free Trade Accord, 1998); (2) Multilateral (the South Asian Preferential Trade Area; South Asian Free Trade Area); and, (3) Sub-regional (the South Asian Growth Quadrangle).

The last strategy, subregional cooperation, provides the broader framework for this paper while the specific focus is on the Kunming Initiative. Initially, the potential for subregional cooperation in South Asia will be discussed briefly then there will be a discussion of the Kunming Initiative whose proposed scope of activity is extended to southern Asian contiguities. A growth quadrangle plan was first mooted at the Male SAARC summit in 1997. Though the proposal eluded consensus at the summit, four out of the seven members – Bangladesh, Bhutan, India, and Nepal – strongly backed the idea of forming the SAGQ. The proponents of SAGQ broadly supported the idea for two reasons: to overcome the political limitations at the SAARC level; and to exploit the potential of contiguous economic zones in South Asia.

Sub-Regionalism

Sub-regionalism is geo-economic derivative of a broadly defined region. Geographical contiguity, resource complementarity, economic viability and, of course, political harmony constitute the basic variables of a sub-regional zone. Even cultural affinity may not be ruled out. A sub-regional zone need not be a microcosm of the larger region. Such an arrangement is a cooperative strategy for joint socio-economic development of the adjoining regions/states. While the broader objectives of collective growth, mutual help and interdependence serve as a common philosophy, local resources, markets, labour and infrastructure rank as the key components of subregional planning. The concept is of Asian origin and has been experimented with at varying levels of application in the Asia-Pacific region.

The potential for sub-regional cooperation as a strategy for economic development in South Asia is well recognized, intra-regionally and by multilateral agencies (ADB, ESCAF and the World Bank). Regions adjoining the northeastern part of the Indian sub-continent – Bhutan, Bangladesh, Nepal and the seven Indian northeastern states – constitute a viable economic zone given their common resources and complementarities. While rich in resources like water, minerals and gas, the region also represents half a billion of the poorest people in the Eastern Himalayan Poverty Triangle. Three major river systems of the region – Brahmaputra, Ganga, and Meghna – constitute a common basin with very considerable but largely unexplored water resources, draining into the Bay of Bengal. No single resource is as rich and as promising as water for the common development of the region. Hydroelectricity, inland waterways, port development and landlocked and transit facilities are the most likely areas for common regional projects. Bilateral agreements have already been reached between India and her three eastern neighbours over water-sharing and power generation. Emphasizing the immense potential for collective development of the region, ADB states that the "dividend of effective cooperation in this sub-region is beyond imagination", especially in the areas of hydropower and infrastructural development.

South Asia is endowed with one of the largest sources of hydroelectric energy in the world. Bangladesh has an estimated hydro power potential of 52000 MW; India, 75400 MW; Nepal, 83290 MW;

Bhutan, 21000 MW; and Pakistan, 38000 MW. However, only about 10% of this energy is commercially exploited (Lama, 2000). Added to the hydro potential are the natural gas reserves whose systematic exploration and exploitation could change the total energy profile of the region. Bangladesh is endowed with the largest gas reserves (13.6 TCP) in the sub-region. A regional power grid for the collective development of hydro and gas resources can be built which can facilitate economies of scale in the production, distribution and marketing of this most vital prerequisite of development in the backward region.

Similarly, the 'seven sisters' of India's northeast have abundant forest resources covering almost half the region with high quality forest species like sal, teak and bamboo, thus supplying the raw material for paper and pulp industries. The region also possesses minerals such as petroleum and natural gas in Assam and the Barak Valley, coal in Upper Assam, Meghalaya, Nagaland and Arunachal Pradesh apart from large limestone reserves. It is in hydro power that the region offers the richest prospects for development as it accounts for about one-fourth of India's hydroelectric power potential. However, poor transport and communication facilities are the biggest handicap in harnessing these resources. A development plan for the northeast therefore has to base "itself upon fuller utilisation of natural and human resources for the purpose of which the economic infrastructure comprising such areas as transport, communication and generation of industrial power from petroleum and hydro sources has to be created . . ." (Thakur, 1999, 114). Calcutta Port is the only developed port in the region catering to the maritime needs of landlocked Himalayan countries and India's northeastern states. A regional transport strategy for South Asia can broadly be developed on following lines:

i) Building the infrastructure of the Chittagong and Mongla ports in Bangladesh thereby relieving the pressure on Calcutta Port.
ii) Development of inland waterways in Bangladesh and Nepal.
iii) Provision of transit routes and access to seaports for landlocked states.

Given the fluidity of the political climate in South Asia and its inevitable impact on SAARC, as amply demonstrated by the postponement of the Khatmandu summit following the Indo-Pak confrontation in Kargil, sub-regionalism is a better alternative strategy to forge regional economies instead of pinning all hopes on the umbrella organization. Rather, the trend in South Asia is already in such a direction. Apart from SAGQ, a further initiative was taken by tying the three South Asian countries skirting the Bay of Bengal – Bangladesh, India, Sri Lanka – with Burma and Thailand into the BIMST-EC (Bangladesh, India, Myanmar, Sri Lanka, Thailand-Economic Cooperation). This latest arrangement is unique in the sense that, for the first time, it has linked three South Asian countries into a common association with two of the contiguous Southeast Asian states.

The Kunming Initiative

At Kunming, the capital of Yunnan Province in China's southwest, an international conference attended by representatives of four adjoining states – China, India, Bangladesh and Burma – was held between August 15-17, 1999. It was a Track II exercise sponsored by the Yunnan Academy of Social Sciences and the Yunnan Provincial Economic and Technological Research Center. There was broad consensus on the need for regional cooperation among the four countries and their sub-regions and the conference decided to establish a Forum for Regional Economic Cooperation with the following objective.

Regional cooperation should be guided by the Five Principles of Peaceful Coexistence, emphasizing equality and mutual benefit, sustainable development, comparative advantages, adoption of international standards, and infrastructure development in order to enhance connectivity and facilitate the widest possible economic cooperation. (The Kunming Declaration, 1999).

The Conference agreed to form a working group representing the four states, which would appraise the respective governments, business and academic communities about the Initiative and prepare areas of priority for promoting regional cooperation among the states concerned. A second conference was held in New Delhi, under the auspices of Centre for Policy Research, in December 2000. Though a supposedly unofficial gathering the Kunming move is officially backed by Beijing while the participation by Burmese delegates in any case represented the military regime. Only Indian and Bangladeshi delegates could truly have participated in the Kunming debates in a private capacity. The hallmark of this sub-regional move simply lies in its brevity. If its economic content admittedly promises great potential, in political terms such a move could be very challenging since the very idea of locking India and China into a single forum will naturally raise more political than economic questions.

The Economic Case

In economic terms, the Kunming Initiative is an ideal plan in that it can interlock the backward northeastern region of India and the equally underdeveloped economies of Bangladesh and Burma with the developed southwestern region of China. These underdeveloped areas, in general, are resource rich but industrially backward and have very poor transport and communication infrastructure. There are considerable manpower resources but the region lacks technical and professional skills. A poor transport system is the bane of the region and the key to its development is systematic planning and large capital investment to construct a network of roads across the tough terrain. In contrast, the Chinese have developed a very sophisticated infrastructure in road building and a communication network. However, the Yunnan region, which is an extension of the Tibean Plateau, contains difficult mountainous terrain straddled by snow-capped peaks and steep valleys. Western Yunnan is traversed by north-south ranges between which the Mekong, Salween and Irrawaddy rivers flow, and these ranges constituted the great barrier between China and Burma in the past, which forced the Great Kublai Khan to lose about half of his forces in invading Burma. And yet, an old trade route came into existence from Kunming westward via Tali and Tengyueh to Bhamo in Burma through which Chinese merchants carried their regular commercial intercourse, trekking for about three weeks to reach the destination, that is, Bhamo. Chinese sources point out that, over 2000 thousand years ago, Yunnan was the overland bridge between India and China, and that, according to B.G.Verghese, even a southwest silk route existed as far back as the second century BC between the two countries.

The great commercial significance of the above route speaks of the wealth of Yunnan province, which was described as one of the richest of the Chinese empire, endowed with valuable mineral and metal deposits, leading to the manufacture of a great variety of goods being exported to Burma. The advantage of connecting the Kunming-Bhamo trade route to India's northeast was emphasised by the British more than a century and a half ago. Captain R. B. Pemberton's classic work, *The Eastern Frontier of India*, published first in 1835, records:

> . . . we have every reason to hope, that if the attempt be judiciously made, a flourishing branch of trade, which is now carried on between its (Yunnan's) industrious inhabitants, and those of the northern Shan provinces of Ava (Burma), may be extended across the Patkoee pass into the valley of Assam; we know that the whole continent of Asia, from Pekin to Cashgar and Yarkund, is crossed by Chinese merchants, in search of a market for their superabundant produce; and we have every reason to believe, that they will

cordially co-operate in any plan which may be suggested to effect this object . . . (Pemberton, 1998).

It would appear as if the Kunming concept was conceived out of Pemberton's idea of more than a century and half ago. As a mineral rich area, Yunnan province has emerged as an industrial hub in China's southwest. Its coal and good quality iron have contributed since the October Revolution to the growth of a large iron and steel industry which is concentrated around Kunming. Much of Yunnan's wealth is concentrated in its copper, marble, phosphorus, tin and zinc deposits and about 90% of China's tin during the 1960s was supplied from this area. Kunming City is a manufacturing centre for cars, cement, chemicals, food products, textiles and other consumer goods. The region's economic activity is destined to grow in greater proportion since China announced preferential investment policies and higher investments in 1992 under its "West Development" strategy. Ten national border trade posts and ten provincial border trade posts have been approved by the central government for the region. Around 200 foreign ventures and agencies have registered with the Yunnan Advanced Industrial Park, in addition to hundreds of foreign-funded business enterprises. Yunnan thus, as a Chinese scholar at the Yunnan Academy of Social Sciences hopes, "will enter the new century with a new image welcoming investment and economic trade cooperation with India, Burma and Bangladesh" (Ren Jia, 2000). It is such rapid economic activity in the immediate neighbourhood to the east that should, therefore, attract India's attention.

Trade and transport are the inseparable constituents of a growing economy. As the Chinese economy has made great strides under a liberal climate, internal and external transport networks were developed at great speed to meet the pressure of growing economic intercourse. Old trade routes are improvised and new transport lines are established. Thus, an industrial boom in the west and southwest compelled the restoration of the old Sino-Burma trade route not just to develop links with Burma, but to gain a gateway to the sea on the west. The Burmese railroads (Myitkyina and Lahio) in Upper Burma are identified as easy transit points to the Indian Ocean and the Chinese plan to build a land route from Yunnan to the Burmese towns of Bhamo and Myitkyina. Once these plans are completed, then the region is surely going to be a busy commercial lifeline between western China and the Indian Ocean, a prospect which India cannot ignore.

Past ties, geographical contiguity and current global trends are all in favour of India linking the destiny of the northeastern states to the Sino-Burmese transit route to the Indian Ocean. Manipur serves as the "entrepôt" to such a linkage as it is situated, to quote Pemberton again, "in the very heart of the great chain of mountains which separate the British and Burmese territories, (hence) this valley (Manipur) presents a most convenient point of assemblage for the traders on either side". A beginning in that direction has already been made in terms of the construction of a 154 km road by the Indian government from Moreh in Manipur to Kalewa on the Chindwin river in Burma, and it is suggested that a bridge across the Chindwin river can connect the Indian side to the Mandalay railway line in Burma. The RITES is undertaking a study on cargo shipping between Calcutta Port and Burma for a final destination to the North Eastern states by road. If such a plan materializes these states will gain sea access. Alternatively, Assam could be connected by restoring the old Stilwell Road from Ledo in Assam to the Myitkyina railroad. The Asian Highway and Trans-Asian Railway sponsored by ESCAP for developing the land transport infrastructure in the SAARC region, have been under construction, though slowly, since the early 1960s and these projects on completion "would not only cater to the surface trade within the region, but also link the region with Iran and Afghanistan on the western side and Burma and Vietnam on the other (eastern) side" (Waqif, 1999, 39). The potential importance of the Asian Highway for playing a crucial role in the northeastern economy is underscored by Rehman Sobhan and Rahamatullah as follows:

> For NorthEast India, as with Yunnan, the AH (Asian Highway) is their only outlet to the East through Burma and onwards to Yunnan and Thailand. Similarly, its land links through the AH with Bangladesh would open up a new logistical environment for this landlocked region. As it stands, virtually all of its export traffic to Calcutta would be diverted to Chittagong port, which remains, by any calculation, the most cost-effective outlet to the sea for this landlocked region (Sobhan and Rahmatullah, 2000).

Integrating the above northeastern routes with the ESCAP network may not be a remote possibility. For example, Asian Highway 1 (A1) entering Bangladesh from West Bengal at Jessore passes through Dacca and Chittagong and then Burma via Technaf, and, once in Burma, it should not be difficult to access it either to the Sino-Burma or the Stilwell roads mentioned above. Similarly, the A2 enters Bangladesh from India at Dinajpur, cuts through Bangladesh into Burma and enters Hochimin City in Vietnam. It is interesting to note that to meet the huge anticipated cargo demand created by the Trans-Asian Highway (South Corridor) on its completion, the Calcutta Port Trust (CPT) Board has already approved development of all-weather lighterage facilities at Saugor Island and a full-fledged port facility with rail-road connections (Roy and

Chakrabarthy, 2001). As a matter of fact, India and Burma, sharing 1600 km of border, formally agreed to develop "infrastructure projects" as announced during the Burmese home minister's visit to India in November 2000. In New Delhi's view, connecting Burma with strong roadways is not only meant to promote mutual cooperation in economic and security areas but also because Burma is seen as a land bridge linking India with southeast Asia and Indo-China. Integrating the Indo-Burmese land routes with the Asian Highways is part of India's long-term strategy to serve its East Asian drive. It is interesting to note that an India-ASEAN plan to launch the Ganga-Mekong Suvannaphoom linkage project at the Laotian capital of Vientiane was announced when the Burmese home minister was still in India. The objective of the project is to promote cooperation in the tourism, education, transport and infrastructure sectors between India and the Mekong Basin countries of Burma, Thailand, Cambodia, Laos and Vietnam (The Hindu, 2000). Given the relative backwardness of the Mekong Basin, it is hoped that India can assist the countries in the region in their development projects in IT, railways, roads, science and technology.

As if illustrative of the commonality of interest, China like India is keen to participate in the development of the Mekong Basin. Yunnan, which is the source of the Mekong headwaters, assumes a key role here too as China and ASEAN are seriously considering the Kunming-Singapore railway line. In fact, the fifth annual China-ASEAN meeting was held at Kunming in April 1999. Kunming thus provides a common hub for both India and China to serve their active involvement in the Greater Mekong sub-region and ASEAN in general, a prospect underscored by the secretary-general of ASEAN in these words: ". . . the southeast Asian countries bordering China and India, as well as other ASEAN members, can deal more confidently and effectively with these large neighbours in the context of ASEAN-China and ASEAN-India relations than on their own" (Severino, 1999).

Speeding up the Asian Highway networks and their integration with the inter-state border routes in the northeast can change the economic profile of this backward region. It is true that road building is humanly and financially an enormous exercise in the northeast with its steep valleys, folding mountains, innumerable river systems and dense forests. Mobilizing construction labour locally is not a big problem and the poor and inhabitants of the region can also find employment. The technical and engineering inputs have to be supplied from other parts of the country but here the Border Roads Organization, which should be more familiar with the topography and logistics of the region, can be involved. The financial commitment will be definitely quite prohibitive, but multilateral funding agencies can be approached either individually or collectively by the concerned governments. ESCAF and UNCTAD are already involved in actively encouraging the SAARC regional transport and communication system. Indeed, the Asian Highway (AH) and the Trans-Asian Highway (TAR) were originally proposed by ESCAF in1959 and 1960 respectively.

India-China Trade

The increasing trade volume between India and China provides another reason in support of the Kunming Initiative. Over the decade since India introduced its liberal trade regime, bilateral trade between India and China has increased nine-fold, though, in terms of volume, it is only a small percentage of their total world trade. However, the potential for increasing the trade volume between the two in the near future cannot be ruled out based on current trends. Now that China is set to join the WTO there is going to be greater scope for increasing the volume of trade between the two neighbours. In fact, under the bilateral trade agreement signed between the two states in February 2000, China has agreed to extend, on a non-reciprocal basis, tariff concessions on several items of substantial value to India, which include rice, fruits, vegetables, marine products and pharmaceuticals, enforceable following Beijing's entry into the WTO. Other instruments of mutual economic cooperation, such as a civil aviation agreement on direct services (1997), an MOU on maritime transport cooperation and a double-taxation avoidance agreement have been signed, though yet to be implemented. Among the products which dominate Sino-Indian trade are organic chemicals, electronic goods, coke and coal, silk and textile yarn from China to India. Furthermore, India exports chromate, alumna, precious stones, tea, food processing machinery, iron ore, marine products, chemicals and castor oil. In fact, Yunnan itself exports about a million tonnes of phosphoric ore to India, apart from phosphoric acid, non-ferrous metals, bristles, palm oil and chilli powder, while the same province imports about 50,000 tonnes of iron ore annually from India (Table 6.1). According to a Chinese source, the total volume of Sino-Indian trade, following economic reforms, increased from \$265 million in 1991 to \$1988 million in 1999. In addition, the volume of trade between Yunnan and India also went up from \$11.34 million in 1966 to \$27.61 million in 1999 (Ren Jia, 2000).

The fear of Chinese goods flooding India through the northeast may discourage India to support the Kunming plan. Such a fear is a real one as P. V. Indiresan observes, but, "while fearing a Chinese take-over of its markets, India forgets that it is equally possible for it to flood Chinese markets. There are many consumer goods that India makes in plenty that are not available to the Chinese people" (Indiresan, 2000). Indian drugs, for instance, have easy access to the Chinese market as do about 20 per cent of Indian textiles. Industrial circles are not much worried about Chinese competition and are confident of facing it except for a few goods. They also feel that the entry of Chinese products may give a competitive push to local industry. A business delegation representing the Federation of Andhra Pradesh Chamber of Commerce and Industry (FAPCCI) which visited China in the summer of 2001 feels that Indian textiles, fabrics, bicycles and pharmaceuticals can compete well with Chinese quality, including even some electronic goods. High quality textiles, manufactured by some of the leading companies may have to put up some resistance, but average consumer products such as sarees, dhotis and tawals have no reason to be concerned. Chinese two wheelers are already in the Indian market and the highly competitive domestic market does not see much of a threat from these imports from the neighbouring state.

Table 6.1. India-China Trade.

Major Indian Exports to China	Major Chinese Exports to India
Iron ore	Organic chemicals
Marine products	Electronic goods
Other yarn, fabrics and made-ups	Coal, coke and briquettes
Drugs, pharmaceuticals and fine chemicals	Medicines and pharmaceuticals
Oil meal	Yarn, fabric and made-ups
Inorganic, organic and agro-chemicals	Machinery (except electronics)
Castor oil	Non-ferrous metals
Processed minerals	Raw silk
Machinery and instruments	Project goods

Source: Ministry of Commerce, Government of India

Sub-Regionalism

Sub-regionalism is premised on devising joint strategies for mutual cooperation in selected areas of proximate geographical zones. Its spatial and sectoral concerns are narrower and do not necessarily embrace all the territorial and economic components of the countries joining together. If any country is worried about the competitive impact of the stronger partner there are enough safety mechanisms to ensure a level-playing field and it also depends on the kind of trade arrangement between the member countries, whether it be a preferential trade area, a free trade area or a loose framework with some tariff concessions and border permits. Wide gaps in major economic indicators like GDP, industrial growth, human development, foreign trade volume and foreign direct investment between India and China notwithstanding, there is mutual complementarity in goods produced. Seen in a larger perspective, as the two largest individual economies outside the OECD, two big emerging markets as well as the most populated societies, and, of course, as two giant neighbours, Sino-Indian cooperation in building cross-border transport and communication infrastructure which can serve their economies is easily justifiable on any economic grounds. The logic of global order shifting from geopolitics to geo-economics alone is sufficient to tie the economies of the two on a firmer and long-term basis. However, transport is the key to trade, a point strongly underlined by Planning Commission member, K. Venkata-subramanian:

> The first priority should be given to transport infrastructure. Sino-Indian trade relations will depend heavily on the construction of trade routes including China-Myanmar-India highways and railways. The two countries also need to jointly establish a sub-regional cooperation community. This proposed cooperation community could embrace China's Yunnan province, north Myanmar, Bangladesh and eastern and northern India. By constructing transport lines, all the big cities in these regions will be linked, and the regions will eventually be turned into a unified growth area (Venkata-subramanian, 2000).

A sub-regional area well served by roads and highways can also minimize the costs and delays in the transhipment of goods.

That the Chinese have emerged economically supreme is rarely disputed, and, with economic strength, goes the power to influence international trade and commercial policies. As a member of APEC, EAEC and ASEM, Beijing's economic clout in the Asia-Pacific region is considerable. The Chinese have already signed several bilateral treaties with Southeast Asian states along with an "ASEAN-China Cooperation towards 21st Century" declaration in 1997. In contrast, India belongs to none of the multilateral economic forums in the region and efforts to gain membership in them have been quite disappointing. Only recently, India was elevated to the ASEAN Summit. Sharing membership with China in a regional group as the one under consideration may improve India's prospects to engage more closely with the Asia-Pacific region. Subregional economic zones elsewhere testify to such complementarity as several adjoining zones of different countries forged close relations through trade and investment. As has been observed, the most rapidly expanding intra-East Asian bilateral trading relationships, China and

Taiwan and China and Korea, have developed around and despite discriminatory restrictions on bilateral trade (Lawrence, 1996).

The Political Case

Politics is at the core of regional cooperation. States coalesce into a regional group to meet a collective threat from a common adversary or to reduce external economic and political dependence. Regional asymmetry, security perceptions, border disputes, external political interference, irredentism and cross-border infiltrations influence the evolution and functional dynamics of a regional economic group. The raison d'être and character of a regional venture is historically conditioned and determined by regional specificity. There can be no blueprint for regional cooperation with universal application to every region, and this applies both to the developed world as well as the developing world. Europe established a Common Market, LAFTA a Free Trade Area, while the ECO neither. The Treaty of Rome (1957) which created the European Economic Community (EEC), known today as the European Union (EU), was intrinsically political in its objective. The war-torn Western European countries conceived the idea to stand together against the common enemy to the east as well as to assert their European identity in a superpower-dominated continent. ASEAN was a post-Vietnam response to ensure the region against perceived Chinese domination. Similarly, regional political equations either sustained or weakened a regional group. The Franco-German axis has been the key to the EEC's successful evolution just as Indonesia's sobering of her regional ambitions enabled ASEAN cohesion, while, conversely, the Indo-Pakistan rivalry is holding SAARC hostage. Kunming, too, is no exception to regional politics.

Kunming's success depends on Sino-Indian consensus. Though China's official support for the Kunming Initiative is in little doubt, India has not yet officially welcomed the idea. However, the idea appears to have drawn the attention of the Indian president, K. R. Narayanan, who during his visit to China in June 2000, met Chinese scholars at the Yunnan Academy of Social Sciences (Kunming) and "highly appraised the importance of regional economic cooperation among the four countries and (the) Kunming Initiative" (Shengda, 2000). India and China are asymmetrical politically and economically, or, rather, in many other respects, including size and population. Regionally, India has shared membership with lesser powers in SAARC and BIMST-EC and in broader organizations like IOR-ARC has assumed a leading role. Within South Asia, which it conceives as its security zone, India has not favourably viewed external involvement nor any lesser power like Nepal or Sri Lanka drawing on external support, a strategic framework which was pursued since independence and forcefully underlined by the so-called Indira Doctrine. Such logic is perfectly understandable within a regional matrix of power. However, is such logic commensurate with a declared policy of economic growth, trade expansionism and investment drive? Will a partnership with China in a regional quadrangle undermine India's role as a regional power or will China gain any extra political leverage through sub-regional participation to check India's influence over the other smaller partners? Burma is close to China, as recent Chinese political support to the military junta and its economic and military assistance to her suggest, and, given the Chinese stake in that country, China may continue to keep the Burmese regime in good humour. However, India, too, has been improving relations with Burma and recognizes her as a key to the Eastern drive. Burma and India share BIMST-EC membership. A 154 km road linking Moreh in Manipur with the Kalewa town of Burma was recently upgraded with Indian assistance. Beijing also cannot influence Bangladesh to counterpose India as Dacca is too involved with India through bilateral and multilateral arrangements, and, given the geopolitical and cultural dynamics of Indo-Bangladesh relations, Dacca would be too constrained to identify with China against India.

China is rapidly gaining access into the Indian Ocean through Burma, as noted above, and such moves are systematic and long-term to support her growing dependence on maritime trade and resources. The Chinese are no strangers to this part of Asia, as they maintained maritime links in the past with some of the remote corners of the Indian Ocean. The Middle Kingdom had developed close contacts with the littorals of this ocean and the Ming dynasty in fact established supremacy over southern Asia up to Malacca and Indonesia, and a rather "shadowy Pax Sinica covered the whole of this area" (Panikkar, 1953, 52). Active trade relations were conducted with Sri Lanka, Kochchi (Cochin), Aden, Madagascar, and East Africa. The post-Ming period witnessed lesser Chinese involvement in the Indian Ocean, which lasted till recently. The 'closed door' policy of later regimes, followed by Mao's greater preoccupation with domestic reconstruction, made China virtually neglect its maritime defence and interests. During the 1980s, China revived its naval interest. The transformation in the structure of international politics and the ideological shift from revolutionary socialism to market reforms at home led her attention to the sea. The importance of safeguarding maritime security, the exploitation of marine resources and the upgrading of naval power came to be widely recognized by the post Mao-leadership, which was powerfully echoed in Deng Xiao Ping's call: "Build a powerful navy with the capability of fighting modern war".

The transformation in the Indian Ocean's strategic climate also explains renewed Chinese interest. The naval withdrawal of the former Soviet Union, America's maritime dominance and India's policy of gaining 'sea control' through a blue water naval strategy could not possibly leave an Asian giant blind to the strategic dynamics in waters not far removed from its western frontier. Maritime goodwill missions, hence, are conducted by the People's Liberation Army-Navy (PLAN) to the South Asian, Arab and African coastal states. In 1986, the PLAN's flotilla sailed around Bangladesh, Pakistan and Sri Lanka. The "Indian Ocean is not India's ocean" has become a frequent refrain in official comments, as were expressed, for instance, by PLAN's visiting commander to Chittagong port in 1986 (Swaran Singh, 1997). It is around the Burmese coast that the Chinese naval presence acquired greater visibility by building strategic bases on Ramree and Coco islands presumably to keep surveillance over Indian naval activities in the region.

These developments obviously caused some security concern in India and the Chinese moves in the Andamans are keenly observed by the Indian strategic community. Against this background, the Kunming Initiative, floated with the official blessing of the Chinese and Burmese governments, cannot elude Indian misgivings. However, within India, opinion is divided on the assessment of the Chinese role in the Andamans, with the opposite school holding that Indian apprehensions as overstated. The former navy chief, Admiral Nadkarni, opined at a seminar in Poona that the Indian navy was making much fuss about the Chinese presence in the Indian Ocean and that it went more often to Southeast Asia and Japan than the PLAN's visit to the west. Sharing such a view was another admiral, who felt that "none of the missions of Chinese navy are, as yet, directly relevant to the Indian Ocean" (Tandon, 1998). It would, therefore, appear that Sino-Burmese strategic cooperation around the Andamans does not deserve to be a weighty factor to dissuade the Indian government from treating the Kunming idea seriously. India has also recently established the fourth command in the Andamans, with the obvious objective of countering a future offensive from the Chinese, or any other country for that matter.

Granting that the Chinese moves are of alarming magnitude, India can still favourably view the Kunming plan. Security and economic cooperation need not be mutually exclusive and in the growth-led and investment-driven climate of the current global economic order, states are compelled to forge economic linkages beyond the immediate security concerns or delink one from another. The logic of economic primacy and the need for greater economic space are temporising traditional rivalries and border disputes. In fact, as has been pointed out, in maintaining the case for Kunming, "stronger economic ties, once established, can themselves be the building blocks for increasing understanding at political, cultural and social levels, which in turn give impetus to further development and consolidation of such economic ties" (Sobhan and Rahmatullah, 2000). States need to realize the imperatives of integrating the 'economics of geography' with the 'economics of development' to adjust to the phenomenon of the regional concentration of trade, growth and prosperity in the world economic system (Panchamukhi, 1994, 19).

Within Asia, countries with long histories of ideological and political conflict are integrating their economies into regional and sub-regional arrangements. Erstwhile rivals of ASEAN – Vietnam, Cambodia and Laos – have bypassed history to join the group. A typical subregional effort comparable to the Kunming Initiative is the Tumen River Area Development programme comprising the two rival Koreas, Japan, China, Russia and Mangolia. It was a $30 billion trade and transport project sponsored by the UNDP in 1992 to transform the backward region which has timber, minerals, oil, coal and cheap labour. The project, whose progress will be conditioned by the course of Korean politics, is significant in its "attention to economic development and reform, a greater willingness by the divided states to be flexible concerning representation, and a closer association between nominally communist states and their capitalist neighbours" (Fawcett and Hurrell, 1995, 244). India can welcome Kunming in this spirit. Looking at China always in an adversarial fashion could deny India even those benefits, wherever feasible. Adopting this line of argument, C.V. Ranganathan, India's former ambassador to China, and a strong advocate of the Kunming idea, suggests:

> Genuine concerns of India which need to be addressed are sometimes conflated with subjective and exaggerated fears resulting in conclusions which favour the continuation of unsatisfactory status quo (with China). This in turn has negative effects on advancing Indian interests, where feasible. Possibly one way to tackle this would be to address realistically and sensitively some genuine concerns of all the concerned parties involved in the promotion of quadrilateral regional cooperation (Ranganathan, 2001, 123).

The Kunming Initiative, therefore, deserves to be viewed in a broader perspective. If seriously pursued, its success may spill over into adjoining areas and groups. Contemporary regionalism is a web of concentric circles with each group representing overlapping membership and mechanisms. Gone is the era of region in its geographical sense serving as the basic unit of regional cooperation. Today's regional blocs are transregional and multilayered and this phenomenon is more relevant to the Asia-Pacific than to other regions. The Chinese are keen about forging economic links with ASEAN states by way of regional cooperation, a theme which is frequently mentioned in Chinese statements. Sun Shihai of Institute of the Asia-Pacific Studies Centre, Beijing, at an Indian Ocean conference in Islamabad, underscored the importance of regional cooperation:

> Many of the countries around the Indian Ocean have realized that regional cooperation has become the mainstream of economic development in the world today. Organizing multi-layer economic cooperation in the region will rationalize the use of the region's resources, work-force, capital and technology, and what is more, raise their collective capability to meet the challenge of western economic regionalism and strengthen their position in their dealings with the developed countries (Sun Shihai, 1995, 89).

Regional cooperation, in Shihai's view, promotes regional peace and stability from a medium and a long-range view, and it should not only cover the economic, trade, social and cultural fields, but must also include political and security matters. Apart from the larger ASEAN, EAEC, APEC and IOR-ARC dominating the southeast Asian region, there are several sub-regional groups/growth triangles constituted on the principle of a natural economic zone (NEZ), which include: the Northern Triangle consisting of Sumatra, northern West Malaysia and southern Thailand; SIJORI of Singapore, Johor and the Riau archipelago, and the East ASEAN Growth area of Brunei, Darussalam, eastern Indonesia, east Malaysia and southern Philippines. In addition to these groups, the Philippines has proposed a new growth polygon comprising Vietnam, Laos, Cambodia, southeastern Thailand, and central and northwestern Philippines. These arrangements, with their growth strategies and free trade regimes, will serve as building blocks to Asia's much prophesied destiny toward a 'new economic power centre' of the world. Kunming can also be one such block, and India's membership in it can complement its vigorous 'Look East' strategy, along with the BIMST-EC initiative.

Conclusion

South Asia is a latecomer to the global family of regional economic groups. While major postwar movements by the third world such as the Afro-Asian Conference (1947), NAM and NIEO were inspired by this region, it lagged behind the African, Southeast Asian and Latin American fraternity in strengthening the process of regional cooperation. When such a process, however, was set in motion in the 1980s, the resulting SAARC remained captive to the Indo-Pakistan conflict. No doubt, following changes in the global political and economic structure, SAARC gathered some momentum, agreeing to establish preferential and free trade areas whose progress, though modest, continues to be marred by regional politics. In other words, South Asia, despite the structural reforms introduced by its states, is far from establishing its identity as a regional collective that could be taken seriously by other dynamic regions. The eagerness of India and her smaller neighbours, such as Bangladesh and Sri Lanka to join the Asian Century, does not enjoy matching enthusiasm from the region east of Malacca. As already noted none of the South Asian states enjoy fully-fledged membership of the economic forums in the East, nor are they likely to be accepted in the near future. Hence, forging links through bilateral and sub-regional strategies remains the best possible alternative to building synergies with the Asia-Pacific. It is here that India can play a leading role in cementing bridges with the East, and Kunming clearly offers one such opportunity.

References

Fawcett, L. and Hurrell, A. (1995), *Regionalism in World Politics* (Oxford University Press: Oxford).

Indiresan, P. V. (2000), 'The Kunming Initiative', *Frontline*, April 14. Indiresan persuasively argues Indian support for the idea, dismissing systematically the MEA's misgivings.

Lama, Mahendra P. (2000), "Designing Economic Confidence Building Measures: Role of India in South Asia", *India's Pivotal Role in South Asia* (CASAC: New Delhi).

Lawrence, Robert Z., ed. (1996), *Regionalism, Multilateralism, and Deeper Integration* (The Brookings Institution: Washington, D.C.).

Panchamukhi, V. R. (1994), 'Recent Developments in Trade Theory and Practice', *DEVINSA*, vol.7, n.1-2, 1994.

Panikkar, K. M. (1953), *Asia and the Western Dominance* (George Allen and Unwin: London).

Pemberton, R. B. (1998), *The Eastern Frontier of India* (Mittal Publications: New Delhi).

Ranganathan, C. V. (2001), 'The Kunming Initiative', *South Asian Survey*, vol.8, no.1, January-June 2001.

Ren Jia (2000), "Promote Regional Economic and Trade Cooperation Among China, India, Myanmar and Bangladesh", a paper presented at the *2nd Conference on Regional Cooperation among Bangladesh, China, India and Myanmar*, New Delhi, 2000.

Roy, H. P. and Chakrabarthy S. N. (2001), "Calcutta Port: Forging Synergic Ties with South East Asia", Paper presented at the *SIOS Conference on Indian Ocean*, Calcutta, February.

Severino. R. C. (1999), *ASEAN Rises to the Challenge* (The ASEAN Secretariat: Jakarta)

Shengda, H. (2000), "Promotion of Regional Cooperation among China,

India, Myanmar and Bangladesh", a paper presented at the *2nd Conference on Regional Cooperation among Bangladesh, China, India and Myanmar*, Centre for Policy Research, New Delhi, December 2000.

Singh, Swaran (1997), "China's Changing Maritime Strategy", *Journal of Indian Ocean Studies,* vol.5, No.1, November.

Sobhan Rehman and Rahmatullah, M (2000), "The Asian Highway and the Trans-Asian Railway in the Context of Regional Cooperation", a paper presented at the *2nd Bangladesh-China-India-Myanmar Conference on Regional Cooperation,* New Delhi, 2000.

Sun Shihai (1995), "Chinese Perceptions on Indian Ocean Security", *Indian Ocean: Security And Stability In the Post-Cold War Era* (Directorate of Naval Education Services, Islamabad).

Tandon, Tandon (1998), *India's Emerging Security Environment with Special Reference to Indian Ocean and its Littorals,* Unpublished Ph.D thesis submitted to Osmania University.

Thakur, Pankaj, ed. (1991), *Profile of Development Strategy for India's NorthEast* (Span Publications Ltd: Guwahati).

The Hindi (2000), Article by K. Venkatasubramanian, "Sino-Indian economic ties", October 25.

The Kunming Declaration, Kunming, China, August 17, 1999.

Waqif Arif A., ed. (1999), *Transport and Communication Development in South Asia,* (Friedrich-Ebert-Stiftung: New Delhi).

CHAPTER 7

South Africa, the Indian Ocean and SADC

Aparajita Biswas

Introduction

Since the late 1980s, momentous changes have taken place in the Southern African region. These changes can be described as turbulent and tragic but also hopeful and positive while impacting both positively and negatively on the lives of individuals, groups, societies, government and states. Nevertheless, in this transitional phase, questions of regional insecurity, peace, collective security and regionalism retain their outmost relevance and have been widely discussed in different forums in what is now often referred to as a 'new South Africa' and/or a 'post-apartheid South Africa'.

With the end of apartheid, when the African National Congress (ANC) came into power in 1994, South Africa joined several regional and international organizations. The ANC felt an urgent need to play an improved and more positive international role that would ease relations between South Africa and the other states around the world. It also felt the need to restructure its pattern of interaction with Africa in general and the Southern African region, in particular. In order to change apartheid South Africa's image as a pariah state and its reputation of aggressiveness to its neighbours and the rest of Africa, the ANC produced a working document that detailed its abhorrence of foreign policy with any hegemonic intentions towards the Southern African region. Instead, the new strategic document advocated regional economic integration to address economic development issues affecting the region. South Africa's decision to join the SADC was a commendable decision, in this context.

Not only in South Africa, and, especially after the end of the Cold Wars, regional cooperation and integration have become attractive policies in most developed and developing states. This is mainly because the characteristics of new regionalism have changed. New regionalism is no more dominated by the Cold War structure, where nation-states were the main actors. The new regionalism needs to relate to the current transformation of the world in which globalisation is the main challenge.

New Regionalism and South Africa

The focus of new regionalism now appears to be on the political goals of establishing regional coherence and identity. As Hettne and others have argued, several features distinguish the "new regionalism" from the old: current processes of regionalism occur more from below and 'within' than before, while not only economic but also ecological and security imperatives push states and communities towards cooperation within new types of regional frameworks. The actors pushing new regionalism are also more varied, including both states and non-state institutions, organizations and movements. Above all, a defining characteristic is that it takes place in a multipolar global context (compared to the bipolarity of the Cold War world), making it extroverted and open, which is one way of coping with today's global economy (Hettne et al; Rao, 2001). Even in regional security matters, regional organizations are playing an active role. The Economic Community of West African States' (ECOWAS) action in Liberia's devastating Civil War, ASEAN's involvement in the Combodian peace process and Russia's attempt to establish a buffer zone in the territory of South Ossetra are but a few examples (Funabashi, 1993, 81).

In the case of the Asian continent, one notices the new regional consciousness which has emerged due to changes in traditional security attitudes and frameworks. There are new diplomatic ties amongst the countries in the region. Thus, since 1990, we have seen the countries such as Russia, South Korea, China and South Korea which have established diplomatic ties with Indonesia, Singapore and Vietnam. India has also strengthened its diplomatic ties with South-East Asia, while Japan is providing financial help to Russia. All of these new developments seem to imply an emerging consensus among Asian states on the eventual disengagement of the US military in the region (Funabashi, 1993, 82).

In Africa, so far, the record of regional integration has been a sobering one, and many regional groupings have been marked by uncoordinated initiation, political conflict and low levels of intra-regional trade. However, some of the external and internal factors that used to impede African integration in the past have somewhat improved in recent years. The launching of the African Union in Durban, South Africa, to end colonialism and unite the people of Africa is a move towards integration. The new organization inherits the OAU's mantle of pan-Africanism, but has a broader mandate to meet the challenges of a rapidly globalizing era. In the inauguration of the AU, Thabo Mbeki, the President of South Africa, declared that "the time has come that Africa must take her rightful place in the world, . . . the time has come to end the marginalization of Africa".

The African Union basically seeks greater African cooperation and integration. To a large extent, the speed of its ratification – that is, within one year of its adoption – makes the African Union unprecedented in African history. It is expected to provide a new launching pad for greater continental cooperation and integration. The lure of regionalism has had profound effects on the foreign policies of African states. This Chapter will examine several sensitive aspects of South Africa's foreign policy, namely its new relation with neighbouring states as a backdrop to the historic changes which have occurred both within the country and outside. These changes call for a radical restructuring of its pattern of integration with Africa in general and the countries of the Southern African region in particular.

Two principal questions have arisen regarding the regional role of South Africa in the post-apartheid period:

1. Since South Africa is economically stable and more powerful than other African countries, would it be interested in cultivating relations with its immediate neighbours, or will it instead choose to develop relations with Europe and Indian Ocean Rim countries?
2. Since South Africa has acquired membership of the SADC, what would be its future role with regard to the Community and vice versa?

The aim of this Chapter is to examine whether SADC will be an appropriate forum to initiate the process of regional economic, political and security integration among the states of the South African region and whether the SADC might constitute a proper agency of non-hegemonic regional cooperation. Before analysing South Africa's relationship with the Southern African states, it is appropriate to profile the sources of South African economic power in the regional context.

Sources of South Africa's Economic Power

South Africa's economy, with a gross domestic product in 1995 of approximately US$120 billion is the largest and most developed in the context of Africa. In regional terms, South Africa is a giant with a population of more than 130 million. It is a giant not just in Southern Africa but in Africa South of Sahara, accounting for 41% of sub-Saharan Africa's gross national product (GNP). Its GNP is 72% greater than that of its nearest rival, Nigeria. Its economic dominance in Southern Africa is widely known. It contributes 78% of the total GNP of the region and its per capita income is three times that of the average of the other SADC states. It is also three and a half times larger than that of the average for the common Market of Eastern and Southern Africa (COMESA) and other SADC counters combined.

South Africa is emerging as a big market in the post-apartheid era after the inclusion of the majority of the population in the mainstream of the country's economy. South Africa possesses very considerable reserves. For example, South Africa accounts for 40 percent of the world's gold and is also the largest producer of platinum, antimony, copper, lead and zinc. The manufacturing sector contributes 24 percent of the country's GDP, although the growth of this sector has been dependent on the continued prosperity and expansion of the mining sector. Thus, despite the spectacular growth of manufacturing industry in South Africa since the World War II, the mining industry has maintained its share of GNP.

Moreover, there was a vibrant link between capitalism in South Africa and international capital (Harshe, 1998, 120-128). In the mining sector, there has been considerable foreign investment. Multinational companies like Anglo-American and old Mutual Sanlaw control the entire mining sector. In fact, in the economy of South Africa, monopoly capitalism has been a dominant feature following the great mineral discoveries in the last quarter of the 19th century. Capital which was organized at the level of the firm very soon took on a monopoly form, resulting in a high degree of capital concentration in the South African economy. This is evident from the fact that Anglo-American companies dominate the economies of South Africa. They are not only large shareholders in South Africa's largest companies, but also conglomerates with investment in many areas particularly in the Southern African region.

There was an obvious connection between the growth of this handful of conglomerates and the dominant role of South Africa in Southern Africa (Harshe, 1999, 87). In fact, these conglomerates controlled the entire mining industry such as diamonds, nickel, copper and coal, in the region. Over the years, the significance of the dominance of South African capital in terms of capital accumulation in parts of the mining sector, such as in the diamond industry, was so much so considerable that diamonds accounted for 72% of the country's exports in the mid 1990s. Not only has the mining sector escalated the country's exports and is a major source of foreign exchange earnings, it proved to be a magnet for attracting a cheap labour force, on a sustainable basis, from neighbouring states. For example, there are around 275,000 legal migrant workers from the Southern African region in South Africa; half are from Lesotho and they provide the country's main source of income, while nearly 60,000 come from Mozambique. South Africa also dominates the transport sector in the region. Its ports and railways handle most imports and exports for Lesotho and Botswana, half for Zambia and Zimbabwe and Swaziland and increasing amounts for Malawi. A single South African company controls most of the forwarding of imports and exports, even those passing through Mozambique.

In the manufacturing sector, foreign capital boosted the process of industrialization, upgraded the existing technology and added to South Africa's military self-sufficiency. In contrast, the neighbouring economies emerged as a kind of monoculture, producing crude minerals and agricultural materials for sale to transnational corporate factories in South Africa and beyond. There was a total neglect of local handicrafts production and there were hardly any efforts made to build up modern consumer goods industries. An inherited institutional structure continued to pressurize national commercial agriculture to focus on profitable exports. It is to be recalled that ever since South Africa was granted independence in 1909, the Southern African region has gradually gravitated towards it. Neighbouring states sought to supply their labour to the mines and plantations of South Africa. Access to labour markets and resources of neighbouring countries contributed significantly to the capital accumulation process in South Africa. This emerging relationship between South Africa and neighbouring countries has been articulated by Immanuel Wallerstein and Seigio Viera in terms of a 'Social Construct', a term not very different to the 'Centre Periphery' concept of the dependency school (Wallerstein and Viera, 1992). For example, not only did states like Botswana, Lesotho and Swaziland belong to a monetary agreement – the Southern African Customs Union (SACU), that dates back in 1910 – but the latter has of late stepped up its capital exports as a result of which South African capital occupies a key position in the economies of most of the majority-ruled states of Southern Africa. As a result, South Africa dominates the economies of these countries. It is the main trading partner for Zimbabwe and Malawi and the main source of imports to Zambia (Lee, 1989).

As a member of SACU, the economies of Botswana, Lesotho and Swaziland are strongly oriented towards external trade with South Africa.

Botswana and Swaziland import 76 percent and 90 percent, respectively, of their goods from South Africa, while the latter accounts for at least 90 percent of Swaziland's imports. South Africa's economic dominance raises the question of the country's ambitions, and, more specifically, the likelihood that it might relish its role as the regional giant and use its position to enhance its own political, diplomatic and economic power. By the late 1980s, "giantism" had developed into one of some ten crises in South Africa's external relations (Venter, 1996), and, by the mid-1990s, Southern Africa reached its 'unipolar moment'. Clearly, pre-1992 SADCC planning, which always endeavoured to draw South Africa into its cooperative net rather than the other way round, was aimed precisely at attenuating this very domination.

However, the relationship between South Africa and the region is one of interdependence. South Africa's reliance on the region is evident from the fact that 25% of all of its exports of manufactured goods are to the region, while South Africa imported primary products in return (Biswas, 1998). On the other hand, South Africa continued to sell predominantly primary products as well as semi-processed goods to the developed industrialized countries, and, in turn, imported various finished products from the latter. For South Africa, neighbouring country's markets are vital for its protected monopolized manufacturing industry. The internal market of South Africa is quite small, and, as its manufactured goods are not competitive, it has to depend upon neighbouring states for the sale of its products. As a result, South Africa's visible exports to the rest of the region exceed imports by a factor of more than 5:1, and the terms of trade always remains in its favour (Biswas, 1998).

South Africa and the Southern African Region

In the post-apartheid period, the accession in August 1994 to the agreement for a Southern African Development Community (SADC) has created a renewed interest in Southern Africa in revitalizing and resuscitating regional groupings. The new government has broadly accepted the proposition that greater involvement in regional trade and projects and obtaining access to regional sectors, could all be of great significance in efforts to promote growth and development in a democratic South Africa. The government emphasis is on a "positive action programme" which could possibly bring about radical qualitative change in South and Southern African relations.

However, post-apartheid South Africa reflects a fragmented division within the state-societal complex. There are apprehensions in some foreign policy circles on the issue of a proper regional role. These differences of opinion came about because of the confluence of domestic and international interests and three contrasting perspectives have been debated. One view – neo-mercantilism – argues that, since South Africa is economically more stable and more powerful than other African countries, it should "promote its own partisan and immediately evident interests, while remaining resistant to the needs and demands of the rest of the region". According to this approach, the rest of Africa is of marginal importance to South Africa's overall development process and thus it would be advantageous to delink itself from any regional organizations, other than those controlled by South Africa, and became attached to one of the other global trading blocs of the world (Davies, 1993).

A second perspective is hegemonistic bilateralism which recognizes the importance of expanding mutually beneficial relations with Africa but without transforming the earlier structural imbalances. On the one hand, it argues for bilateral or multilateral cooperative enterprises over a number of sectors such as transport, industry, energy, health and security, but, on the other hand, it would tend to be rather "lukewarm about making a commitment to any regional programme or a regional organization". This view would envisage the development of a Southern African Common Union (SACU) within which South Africa would play a hegemonic role. This approach is premised on the idea of South Africa as being the regional power (Davies, 1993).

A third perspective projects South Africa as a full and active partner in a regional programme. It emphasizes a positive action programme which would possibly bring about a radical qualitative change in South and Southern African relations. Its objective would be to contribute to the development of an agreed mutually beneficial programme which would need to combine, in an appropriate and realistic way, elements of cooperation, co-ordination and integration rather than being a simplistic polarized choice between co-operation and integration (Davies, 1993).

In addition to these three perspectives, possible futures for "Southern Africa Developmental Cooperation/Integration" and "Ad hoc Cooperation" have been debated. These competing visions for the future of Southern Africa and South African role in the subcontinent has an institutional pedigree. There were two main regional institutions ostensibly promoting regional cooperation/integration in Southern Africa during the transitional period – the Southern African Custom Union (SACU) and the Southern African Development Coordination Conference (SADCC). There was also a third path of informal ad hoc cooperation. Each reflected distinct architectural principles for the institutionalization of regional economic relations in post-apartheid Southern Africa and each promoted a disparate interest within South Africa.

SACU was primarily concerned with trade and possibly monetary cooperation and represents market cooperation/integration where the integration force of the market is released through the removal of restrictions and barriers to regional trade, rather than through positive government intervention. This process favoured the deepening of cooperation by stages from a free trade area to a common market. Thus SACU was based on the classic process of market cooperation where cooperation/integration proceeds along a linear path, with four steps – a free trade area, a customs union, a common market, and, finally, an economic union (Davies, 1993).

SADC counts all ten regional states as members – Botswana, Lesotho, Malawi, Mozambique, Namibia, Swaziland, Tanzania, Zambia, Zimbabwe, and, as of 1994, South Africa. SADC's immediate precursor, the Southern African Development Coordination Conference (SADCC), was the frontline states' response to P. W. Botha's "total national strategy", which included his call for a Constellation of Southern African States (CONSAS). By expanding the common market (SACU), CONSAS would have tied South Africa's neighbours closely to it as the subcontinent's hub of transportation and trade. When South Africa's neighbours rejected CONSAS, and formed SADCC in 1980, Pretoria undertook a massive destabilization programme, which included military forays as far as north as Zambia. SADCC was reconstituted as the SADC in 1990 in anticipation of apartheid's demise and South Africa's future membership. Cooperation under SADC includes a wide range of issue areas and its process is purported to further regional cooperation and integration based on the concept of interdependency. It is predicated on multilateral negotiation in baskets of issues that would facilitate trade-offs in such areas as labour mobility, industrialization, trade, water and energy. Each of these areas is characterized by strong regional interdependencies. This approach comes under the general rubric of development cooperation which was expected to ameliorate the natural inequality that accompanies market integration. It does so through the strong role of the state, contradicting the neo-liberal prescription (Garden, 1999).

These different viewpoints came to fruition because of a confluence in domestic and international interests. In post-Cold War international relations, South Africa has emerged as a middle power owing to the evolution of a distinct pattern of ties between state and society. Prior to the 1994 election, the ANC, in its Research and Development Plan, prescribed a strong state role. It favoured the SADC as the institutional vehicle for promoting development integration. However, after coming to power, it has not only shifted its stand and adopted a firmly neo-liberal economic policy, but has also started moving towards accepting the free trade and liberalization argument. In other words, the post-apartheid situation in South Africa compelled social forces with various ideological convictions to compromise their position (Westhuzian, 1998, 437). This compromising approach at the domestic level also reflects in its foreign policy formulation which has become increasingly aligned with neo-liberal doctrine and President Thabo Mbeki is recognized as one of the driving forces behind this shift. He has been instrumental in establishing a close relationship with Washington via the US-South African Bilateral Commission. He has also been a regular participant in the World Economic Forum and the Partnership Africa Conference where he has sought to attract international capital to the region on the premise that the SADC is "involved in negotiations to transform the subcontinent into a free trade area and further enhance the attractiveness of the region as an investment destination". This rightward shift can be explained from the very fact that South Africa occupies an intermediate position in the international political economy and intends to play an intermediary role between foreign capital and the resources of the neighbouring states which are decidedly peripheral in the international political economy (Westhuzian, 1998).

Once again, South Africa's aspiration to be a middle power has become manifest in its growing interest in international institutions as a focus of its diplomacy. South Africa has declared itself as a non-aligned country, has signed the non-proliferation treaty and has joined a number of inter-governmental organizations and multilateral treaties. South Africa has committed itself heavily to reform the UN, the IMF, and the World Bank. The country's involvement in these institutions shows its willingness to assume a 'middle power leadership' role within the broader global rubric. In other words, in any international common issue, such as the North-South divide, South Africa wants to be a part of multilateral initiatives for improvement in the relationship. Speaking on this issue in a Senate meeting in May 1996, Foreign Minister Alfred Nzo remarked that "South Africa is a developing country with certain attributes of a developed or industrialized country. This enables us to understand, and relate to, the concerns of both the South, as well as the North, and therefore to play a pivotal role in drawing them closer together to promote international developments". South Africa has taken initiatives in proposing the establishment of the Indian Ocean Rim Association along with the other developing countries for regional cooperation and development. In fact, South Africa tries to bridge the gap between developed and developing states and assumes a reformist stance that is closely in line with an emerging foreign policy agenda following its accession to the presidency of UNCTAD (IX) and its Chair of the Non-aligned Movement (NAM).

However, SADC member states are sceptical about the Free Trade Policy and this has created tensions in South and Southern African relations. The conclusion of the European Union-South Africa bilateral free trade agreement of March 1999 does not sit well with many SADC/COMESA governments, who are suspicious of South Africa's hegemonic impulse. As a result of this FTA agreement, the Southern Africa region will be confronted with increased competition from EU goods, both in their own markets and also in the case of industries that produce for the South African market. Even they are also sceptical about South Africa's proposed Indian Ocean Rim Association and for free trade across the Indian Ocean which they fear would jeopardize intra-regional trade if Southern African markets are forced to open to larger and more efficient economies such as Australia, India and Mauritius (Venter, 1996). Against this background, it is worthwhile to consider South Africa's regional dilemmas regarding the Indian Ocean Rim.

South Africa and the Indian Ocean Rim

The concept of Indian Ocean Rim (IOR) was first articulated by Jawaharlal Nehru in 1947, when he first commented on the common origin of the people surrounding the Indian Ocean area. The idea was revived in 1970, when the Indian Ocean Zone of Peace (IOZP) concept was mooted. Unfortunately, concrete groupings of Indian Ocean states were not established for a number of reasons, including the fundamental problem of defining the region. Thus, in spite of the tremendous potential for cooperation among states in the region there are not many regional and sub-regional groupings in the Indian Ocean compared to the Pacific and the Atlantic Oceans. Major sub-regional groupings include the Association for South East Asian Nations (ASEAN), the South Asian Association of Regional Cooperation (SADC) and the Gulf Cooperation Council (GCC). It was in the 1990s that the initiative to form a regional group in the Indian Ocean was taken by seven countries in the region, namely, India, South Africa, Australia, Mauritius, Oman, Singapore and Kenya. The initiative came at the time when the world witnessed significant economic and political shifts, including the end of the Cold War, a growing global trend towards regionalism, and the end of South Africa's international isolation. The other significant developments during this time were the emergence of a liberal international trading environment with the conclusion of the Uruguay Round, the General Agreement on Tariffs and Trade (GATT), the establishment of the WTO and the adoption of market economies by an increasing number of countries, including India.

The idea was floated by the then South African foreign minister, Pik Botha, when he visited India in 1993. In 1995, the concept of the IORI was formally launched when President Nelson Mandela, on an official visit to India, strongly supported the prospect of such an initiative. He spoke about the necessity of the states of the Indian Ocean Rim to come together on a single platform in the present international system. He noted that the natural urge of the facts of the history and geography should broaden itself to include exploring the concept of Indian Ocean Rim and socio-economic cooperation, and other peaceful endeavours (Mandela, 1996).

This proposal was enthusiastically received by both the Indian and Australian Governments. For India, this was the time when it wanted to assert its position in stronger regional groupings and had initiated a 'Look East Policy'. For Australia, this was in line with its 'Look West' strategy announced by the Ministers of Foreign Affairs and Trade in August, 1994. Also, in 1995, the Australian foreign minister had ideas of contesting the UN Secretary General's position and wanted political support for his candidature. For South Africa, the need for regional identity and its aspiration for middle power leadership became manifest in its growing interest in international institutions and organizations as a focus of its diplomacy. Thus, the formation of an Indian Ocean Rim Association was the manifestation of a combination of events that had transpired at that time with respect to regionalism amongst the countries of Indian Ocean.

According to its Charter, the IOR-ARC is foremost an outward-looking forum for economic dialogue and cooperation with the following key objectives: improved market access though trade liberalization, and the facilitation of free and enhanced flows of goods, services and investment throughout the region. Aside from its role as a forum for strengthening trade liberalization along WTO lines, the IOR-ARC "is designed to set directions for the economic and trade policy in the Indian Ocean Region". The IOR-ARC is thus to be firmly based on the principle of 'open regionalism', as encouraged by the WTO (Mills, 1998).

Despite the formulation of the Charter, the problem of the scope and speed of membership accession remains problematic. Regarding the membership issue, from the outset, India was against the idea of including any discussion of strategic issues in the organizational forum because of the potentially divisive nature of such a security debate. The strategy regularly employed by India in its participation in the IOR was to exclude security or strategic debates. This has prevented the forum from being used as an instrument to comment on the Kashmir issue, for example. India was opposed to the inclusion of security matters in IOR-ARC, an issue on which she seriously disagreed with Australia. In the debate, representatives from Australia, South Africa, Indonesia, Singapore favoured a first track approach to the membership of IOR-ARC. Australia, for example, would like to see as many as 35 countries quickly incorporated within IOR-ARC. It takes a "liberal attitude to membership" and believes that IOR-ARC cannot proceed apace without the inclusion of major economies such as India, Pakistan, Bangladesh and Thailand, although it sees problems in extending the definition to Iran and Egypt (Mills, 1998).

Again, on the membership issue, India also opposed the adoption of the 'inclusivity' principle, which enables an open-ended membership along APEC lines. Instead, India preferred a cautious 'building bloc' approach, both to membership and to the inclusion of non-economic issues in the agenda. India's cautious or 'evolutionary approach' to membership was based on the absence of any history of successful cooperation in the IOR region. India's hesitancy also related to its own tensions with Pakistan over Jammu and Kashmir, the Tamil separatist movement in Sri Lanka and the Bangladesh break away from Pakistan. Hence, India prefers a small group of countries taking the lead in first building up norms of cooperation within IOR groupings. The application of Pakistan for membership was rejected by IOR-ARC as it did not offer MFN status to one of the IOR-ARC members, that is, India. Besides the membership issue, there were also opposing views regarding the inclusion of security matters within IOR-ARC's scope. On the one hand, South Africa and India were agreed that they should be committed in order not to jeopardize cooperation in, or 'divert attention' away from the economic sphere, while, on the other hand, the Australian government reportedly pushed very hard to have security included in the agenda. Thus, these two issues of membership and security came into IOR-ARC before substantive issues were sorted out and integration was deepened. According to senior South African officials involved in the negotiations, these two issues "held back" progress in the first year of talks and ultimately "almost scuttled" the whole initiative (Mills, 1998).

Conclusion

From South Africa's perspective, a number of commentators see substantial benefits in terms of greater market access that this Association might facilitate. For example, it has been argued that "Although defining the region may be the initial problem to be overcome, it is a minor one in comparison to others that loom on the horizon for a potential IOR economic grouping" (Garden, 1999). Indeed, the South African Foreign Policy Discussion Document, released in June 1996, noted that "South Africa's involvement in the South Atlantic and Indian Ocean regional groupings merits special attention". The official position of the South

African government is that cooperation with IOR should take place within South Africa's broader commitments to the WTO, and, second, that cooperation will benefit its fellow SADC members economically and ultimately, be to the economic benefit of Africa as a whole ((Marx, 1995, 7).

The South African Foreign Trade Organization (SAFTO) supports the IOR idea on the basis of two principles. It argues that, from South Africa's trade experience with sub-saharan Africa, it is not always sound policy to target continental neighbours as export partners. More explicitly, South Africa's potential regarding partners on the African continent is that they do not provide sufficient common features and cohesion expected in a single viable market (Mills, 1998).

Moreover, geographically, South Africa is the hub of the southern hemisphere, linking South America with Australia. As a result, the IOR seems to offer South Africa an attractive alternative to redefine its target markets. SAFTO identifies the more powerful alliance, including features common to South Africa and the IOR countries. These are: a) an historical heritage of trade and cultural exchanges; b) easy access of land and sea; c) conductive conditions for exchange in agricultural and mining; d) cooperative research in education, telecommunications and the environment, and, above all, e) technology transfer (Mills, 1998).

However, South Africa's enthusiasm in Indian Ocean Rim Association (which excludes SADC) creates stress in relations between the South African government and Southern African countries. From the Southern African point of view, South Africa's approach to the Indian Ocean Rim countries does not take account of South Africa's responsibility to the African continent as a whole, and particularly the Southern African region, and will dilute the government policy thrust on Southern African regional co-operation and development.

It will be difficult to protect local enterprises from competing technologies coming from overseas. These concerns may, to some extent, explain the South African government's current ambivalence towards IOR initiatives, treating them as a matter of general interest. Had South Africa entered IOR-ARC with its SACU neighbours and had an assessment of its membership been positive, it would have been argued that any potential benefits would not only spill over into the Southern African region, but also would promote South-South cooperation. For many, South Africa's decision to join IOR-ARC has far-reaching negative implications for Southern African regional integration (Mills, 1998). The perception has thus been created that South Africa is not committed to regional cooperation.

Thus, overall, South Africa would like to balance its commitment to these various regional organizations. In order to rectify past antagonisms and build bridges with African neighbours, South Africa's focus has been mostly in converting the Organization of African Unity (OAU) to an African Union and this objective is tied to the New Partnership for African Development, initiated in 2001. The African renaissance has become the driving force for new partnerships and IOR-ARC has almost become a peripheral issue. To keep IOR-ARC alive, however, and to achieve its objectives, what member states need is strong determination and political will.

References

Biswas, Aparajita (1998), "The New South Africa in Southern Africa", *Africa Quarterly*, Vol. 38, No. 2.

Biswas, Aparajita (1999), "Sub-regional Cooperation in the Indian Ocean", a paper presented at the African Studies Association Conference in Australia, December.

Davies, Robert (1993), "Emerging South African Perspective on Regional Cooperation and Integration After Apartheid", in Bertel Oden, ed., *Southern Africa After Apartheid*, (Upsala), pp. 78-83.

Funabashi, Yoichi (1993), "The Asianisation of Asia", *Foreign Affairs*, Vol. 72, No. 5, November-December.

Garden, Freer (1999), "South Africa and the Indian Ocean Rim: Between SADC and the Deep Blue Sea", *Indian Ocean Review,* Vol. 12, No. 2, September.

Harshe, Rajen, *Twentieth Century Imperialism: Shifting Contours and Changing Conceptions* (Sage Publications: New Delhi).

Harshe, Rajen (1999), "South Africa and Trajectories of Regional Cooperation in Southern Africa", *Africa Quarterly*, Vol. 39, No. 4.

Hettne, B. et al, "The New Regionalism: A Prologue" in B. Hettne, S. Intoi and O. Sunkel, eds., *National Perspectives on the Next Regionalism in the South* (Macmillan: London).

Lee, M. C. (1989), *SADCC: The Political Economy of Development in Southern Africa* (Winston: Derek Publisher).

Mandela, Nelson (1996), Nelson Mandela, "South Africa's Future Foreign Policy", *Foreign Affairs Discussion Papers*.

Marx, J. (1995), "South African Foreign Policy in the New Era: Priorities in Africa and the Indian Ocean Islands", *South African Journal of International Affairs*, Vol. 2, No. 2, Winter.

Mills, G. (1998), *South Africa and Security Building in the Indian Ocean Rim* (Johannesburg: The South African Institute of International Affairs).

Rao, P. V. (2001), "The Global Context of Regional Co-operation in the Indian Ocean", in P.V. Rao (ed.), *Regional Cooperation in the Indian Ocean* (New Delhi, South Asian Publishers).

Wallerstein, I. And Vieira, S. (1992), "Historical Development of the Region in the context of the Evolving World System", in Sergio Vieira, William Manten and Immanuel Wallerstein, eds., *How Fast the Wind in Southern Africa 1975-2000*, (Trenton, NI: Africa World Press).

Westhuizen, J. van der (1998), "South Africa's emergence as a middle power", in *Third World Quarterly*, Vol. 19, No. 3.

Venter, D. (1996), "Regional Security in Sub-Saharan Africa", *Africa Insight*, Vol. 26, No. 2.

NON-TRADITIONAL SECURITY CHALLENGES

CHAPTER 8

An Agenda for Environmental Security in the Indian Ocean Region

Timothy Doyle

This Chapter will explore the concept of environmental security within the Indian Ocean Region (IOR). An appropriate draft agenda will be discussed using seven different but intertwined examples of environmental security issues taken from states within and bordering the Indian Ocean – land degradation, water, fisheries, climate change, nuclear waste, environmental refugees and the urban explosion and associated environmental deterioration.

Depending on the disciplinary paradigm, whether it be international relations, critical geopolitics, military security, or environmental politics, definitions of environmental security are as numerous as definitions of what constitutes the 'environment' itself. In the recent environmental security literature, issues which fall under its rubric are multifarious and diverse. Some of these include: biological and ecological security; the greening of military operations; climate change; desertification; biodiversity; human population and migration; fisheries; forests; energy; water; nutrition; shelter; and poverty.

Two categories can be used to order this cacophony of issues in the first instance. First, most environmental security issues are still cast around the 'security' of the nation-state. Secondly, a more inclusive definition, one which transcends nation-state boundaries, relates to conditions which secure individual access to a basic infrastructure for survival in a geopolitical region defined by shared environmental boundaries. Environmental security, in this vein, is reliant on shared understandings of ecological conditions leading to potential and real conflicts, as well as developing a more sustained, peaceful, and resource-secure regional future.

What is Environmental Security?

The environment has often been used as a tool of war, from the salting of Carthage to the Russians scorched earth retreats before the armies of Napolean and Hitler. Plato, mocking the notion of a republic of leisure, argued that such a regime would soon resort to a war to satisfy its taste for more space and natural resources. But sustained thinking about the environment-conflict connection is a product only of the last few decades. While clashes over non-renewable resources such as oil or gold are as familiar as the Persian Gulf war, the question now is about the role of renewable resources such as water, fish, forests, and arable land (Dabelko, 1999, 2).

Although there have been conflicts over resources since the earliest human societies, interest in both renewable and non-renewable resources within environmental security frameworks has dramatically increased since the end of the Cold War. Security is usually understood in state-centric terms, "concerned with intentional physical (mainly military) threats to the integrity and independence of the nation-state" (Scrivner, 2002, 184). Immediately after the so-called 'victory of capitalism' and the breakup of the communist-inspired USSR in the late 1980s, world orders which had existed since World War II were called into question. During this time of uncertainly, there emerged a global, almost post-modern, policy-shaping concept embracing a shared plurality of interests which crossed nation-state borders, commonly referred to as *multilateralism*. The multilateralist decade of the 1990s, which ended as the current phase of US unilateralism emerged forcefully in the new millennium, was an era when new boundaries and borders were drawn in the sand, as alternative concepts of identity and collectivity were imagined. One such idea which evolved at this time was that of *environmental security*.

This trend was reinforced and supported by military establishments who sought means by which they could continue to justify Cold War levels of military expenditure during an apparent time of peace and prosperity for the West. As a result, both the US and Russia formed high level units of environmental security within established security institutional infrastructure, such as the Pentagon's Centre for Environmental Security.[1] Paradoxically, perhaps, some peace advocates also championed the concept. In a book appropriately entitled *Green Security or Militarized Environment*, Jyrki Kakonen writes:

Peace researchers have argued for environmental security in order to show that . . . national defence resources could be used for civilian purposes in the field of environmental problems. The aim is to convert military resources . . . to do the environmental protection in order to transform the military into a paramilitary and further into a non-military organization. This is an option after the Cold War, but there is a danger that the militarist approach to deal with environmental issues leads to the militarization of the society . . . (Kakonen, 1994, 4).

As with the now dominant concept of *terrorism*, the *environment* is an extremely powerful multilateral concept which crosses nation-state borders relatively easily. In this vein, a new enemy is imagined, one which was not a human terrorist; but *nature* itself. This interest in a *combative environment* is not new. In western terms, it has been aptly recorded in the 18ᵗʰC works of Thomas Hobbes. In his most famous work – *Leviathan* – Hobbes depicts nature as in a state of perpetual war with itself: ". . . and the blood-dimmed stains shall be loosed upon the world and all anarchy will prevail . . ." (Hobbes, 1772). This conservative view of nature was used by Hobbes to justify his call to create an all-powerful authoritarian 'machine' which would be the only means to avert global environmental catastrophe. Of course, in this understanding of nature, humanity is also in a perpetual "state of warre" (Hobbes). This western understanding of the 'state of nature' is not just restricted to neo-Hobbesians, but has substantial populist credence, as most western imaginings comprehend peace to be the aberration whilst war is construed as the 'natural state'.

This interest in Gaia as the common enemy (as well as a pathway to common salvation) was further heightened in the West due to increasing, but rather late, understandings that the minority world (the more affluent world) had to share its basic survival systems with the majority world (the less affluent world). This concept of a shared spaceship Earth had been vociferously pushed by western environmentalists since the late 1960s, but, due to characteristic conservatism of the disciplines of international relations and security studies this green rhetoric was only picked up in the academic literature and the governmental grey documentation in the late 1980s and early 1990s.

This interest in environmental security emerged forcefully in the Brundtland Report in 1987, and increased at the first Earth Summit in Rio de Janeiro in 1992. The nexus among environment, development and security was never stronger than at the recent 'Earth Summit Plus Ten' in Johannesburg in 2002. The notion of environmental security, however, is hotly contested. Its most common variation is concerned with the impact of environmental stress on societies, which may lead to situations of war within and between societies. In this manner, environmental security agendas are about seeking issues, which, if not addressed, may provide the basis for increasing human conflicts. In this sense, environmental security is understood in somewhat negative terms.

The advantage of this redeclared enemy – nature itself – was, or so it was believed in the Westphalian mindset, that it constituted a 'common' security issue for all humankind, or in the words of Brundtland 'Our Common Future'. Of course, the symbol 'environment' is not common at all; rather, the issues which gather under its umbrella are culturally diverse. Concepts of environment are far from apolitical; rather, they are the exact opposite (Doyle, 2003). They are intensely politicized categories utilized to redraw boundaries of collective identity, behaviour, political activity, security and, most importantly, power and resource distribution. Elsewhere, I write:

> So environmental politics is not just about 'goodies' versus 'baddies'. This symbol environment has such power that numerous cultures, and the powerful and powerless within them, invoke its name for disparate purposes (Doyle and McEachern, 1998, 4).

That the 'environment' flag means many different things to different people, does not make it a charlatan. Rather, it is a symbol almost as broad as nature itself. It is perfectly chosen for the infinity of possible responses which gather under and around it. Just when it appears that a safe net of definition can be cast over it, it wriggles out, and takes on a new guise, in a separate context (Chaturvedi, 1998; Doyle, 2003). While the list of issues which make up an environmental security agenda is an important list, it is an extremely subjective and culturally determined one. Such an agenda must be cognisant of the different meanings and issues and respectful of these differences and variations if practical solutions are to be uncovered which make sense to the people who are most affected by environmental degradation.

At the end of the 1980s, when these agendas for common futures were first being drawn up, predominantly *western* issues were being recast as *global* ones. The minority world, in this light, portrayed the major problems of the globe as species extinction, global climate change, desertification, and overpopulation.[2] Needless to say, at the end of the 1980s, these were not issues high on the environmental agenda as defined by most people living in the majority world. Other issues of more immediate survival dominated (and still do) such as health, shelter, food and water security. In a provocative book, entitled, *Tears of the Crocodile*, it is argued that the developed world has managed to divest itself of its responsibility to the global environment by moving the arena 'away from people and onto things, forces'. The authors write:

In short, the developing world for the first time, is being asked to be an equal partner in a world-wide endeavour precisely because the emphasis has shifted away from the needs of the poor. By advancing an environmental agenda the North has once more concentrated on its own interests and has called them globalism (Moyo et al., 1993, 5).

Within this western framework, when people were seen as part of an 'environmental security' agenda, ambiguously people are not perceived as part of the environment; they are simply users or, in the case of the poor, degraders (Doyle, 1998). In 1990, the United Nations Human Development Report argued that poverty is one of the greatest threats to the environment, and, in 1993, the International Monetary Fund (IMF) announced: "Poverty and the environment are linked in that the poor are more likely to resort to activities that can degrade the environment" (International Monetary Fund, cited in R. Broad, 1994). There are two key problems with this line of argument. First, all poor people are regarded in an homogenous fashion. An important distinction must be made on the connections between *types* of poverty and environmental degradation. For example, those still operating subsistence lifestyles (though under threat); those who have been recently removed from this lifestyle; and those people who have long ago been driven to the precipice of survival (the 'landless and rootless') have very different relationships with their environments. The latter have no security of tenure and little connectedness to place. This category includes those peasants and squatters who survive by cutting forest cover, by consuming wildlife, and by planting crops on soils which will erode.

Secondly, many western environmental security theorists fail to weigh up the costs of advanced industrialism on a global scale; not just within the boundaries of nation-states. Issues of overconsumption in the minority world – and by the minority world – cannot be underestimated. The fact is that the U.S. and Japan together represent 40 per cent of the world's Gross National Product cannot be denied (Imura, 1994). In the Indian Ocean Region (IOR), the consumption patterns of Australians far outweigh most of, for example, their Indian and East African neighbours.

Obviously, this agenda for this kind of common environmental security was greeted with some skepticism in parts of the majority world. Dabelko writes:

In less developed countries of the world, these ideas have elicited mixed emotions. Obtaining food and water is a daily struggle for the world's 800 million malnourished people, and according their problems the high priority of a security issue obviously has great appeal. But leaders in. . . . Cairo, and Kuala Lumpur also fear that such an approach will invite violations of their national sovereignty as outside powers intervene to "help". They gave a frosty reception, for example, to Gorbachev's 1988 proposal to complement the blue-helmeted armed forces serving under the United Nations with a "Green Helmet" force to react to natural catastrophes and environmental crises (Dabelko, 1999, 14-18).

A concept of environmental security which is more inclusive of the interests of the majority of people in Indian Ocean states, both littoral and non-littoral, is one that moves away "from viewing environmental stress as an additional threat within the (traditional) conflictual, statist framework, to placing environmental change at the centre of cooperative models of global security" (Dabelko and Dabelko, 1995, 4). However, to do this, there must be increased understanding of the environment, not as an external enemy force, but as a diverse nature which is inclusive of people; a nature which has the potential to provide secure access to individual citizens of all IOR states to basic nutrition; adequate access to healthy environments; appropriate shelter; and, a security to practice a diverse range of livelihoods which are both culturally and ecologically determined.

Seven Important Environmental Security Issues in the Indian Ocean Region

Whilst many parts of the minority world are currently seeking technological solutions to environmental problems, for many IOR researchers the major task is still documenting the list of environmental problems, attempting to collate base-line data which is sadly missing.

There is an enormous gap in the literature on environmental security and the Indian Ocean region. There have been substantial academic works completed in recent years on the broad subject of environmental security (see, for example, Myers, 1993; Dabelko and Dabelko, 1995; Broda-Bahm, 1999; Lowi and Shaw, 2000; Redclift, 2000; Cheremisinoff, 2002). Some of these works move from theory into empirical research, but when this occurs, most of this scholarship is based in and around the Atlantic, Pacific and Southern Oceans (see Kakonen, 1994; Barnet and Dovers, 2001; Dokken, 2001; Foster, 2001). It has been very rare that researchers have utilized the concept of environmental security in the Indian Ocean region.

This lack of research literature reflects a broader neglect by the more affluent, minority world in addressing social science policy issues confronting the South. There have been some articles addressing a particular environmental security issue in a specific state, such as water wars in the Jordan Basin (see Shaheen, 2000) or environmental degradation leading to human displacement in South Africa (Singh, 1996).

In a search of the electronic version of *Expanded Academic Index*, only one reference emerges which includes environmental security insights into the Indian Ocean region as a whole (Chaturvedi, 1998).

Obviously, establishing an environmental security agenda for the Indian Ocean region is also significant in that it addresses basic survival issues which affect the inhabitants of the region, approximately one third of the globe's population. Chaturvedi writes:

> The Indian Ocean has been rightly described as the 'Heart of the Third World' or the 'Ocean of the South', with low per capita income and low levels of development in the majority of countries. The overwhelming mass of these peoples struggle to survive under the conditions characterized by chronic poverty, precarious political systems, stagnating and struggling economies, fragmented political systems guided by the considerations of ethnic identities . . . (Chaturvedi, 1998, 712).

The concept of environmental security must be brought to life by reference to some of the most pressing environmental issues confronting the Indian Ocean region. Just seven broad areas are identified here, each one ecologically interlocking with the other, snowballing in magnitude, creating desperate realities for billions of people culminating in abject poverty, both in terms of biodiversity (or lack thereof) and in terms of human existence: land degradation; water; fisheries; climate change; nuclear waste; environmental refugees; and urban explosion and deterioration.

Land Degradation/Deforestation

As Graeme Hugo writes, that much "contempory environmental degradation in Less Developed Countries has its real roots in historical processes such as colonial exploitation which produced different modes of agricultural and pastoral activity to meet the needs of the colonial power and different patterns of population growth and distribution from those which prevailed in precolonial times" (Hugo, 1996, 124). This is a very broad environmental security issue, and its causes are many. Obviously, an important contributor is deforestation. In the Indian Ocean region vast areas of forest have been eradicated. The size of these areas simply defy the limits of our imaginations; literally beyond quantification. Even in the more affluent parts of the Indian Ocean region, this decimation has continued unabated despite the concentrated efforts of the Australian environmental movement which, for thirty years, has had a principal focus on forest conservation, sometimes to the detriment of other pressing environmental issues (Doyle, 2000). Australia is currently listed fourth on the list of so-called "third world" deforesters of hardwoods after IOR partner Indonesia, Brazil, and the African Republic of Congo. Satellite imagery has also revealed that less than 15% of South Asia is also currently forested (Hassan, 1991, 13). Deforestation, coupled with poor agricultural practices, and continuing tectonic movements, have led to floods, soil erosion and siltation. In his work on environmental issues and security in South Asia, Shaukat Hassan writes in direct relation to the Ganges river bed, which has risen by over 0.5m in the early 1990s until the present day:

> The result has been increasingly disastrous floods, as well as the heavy siltation of tanks and reservoirs, whose life expectancy has been reduced by a third because several times more sediment flows into them than was calculated at the time of their construction. Among the downstream consequences of soil erosion are the choking of estuaries and harbours, and land formations in the Bay of Bengal, both of which have created conflict between India and Bangladesh (Hassan, 1991, 16).

In Africa, land degradation is severe. Commercial afforestation, logging and grazing have been massive, leading to a loss in biodiversity and increased run-off which removes top soil and increases siltation. In research dedicated to environmental security and displaced people in Southern Africa, Meena Singh explains:

> Soil loss in South Africa has been estimated to be approximately 400 million tonnes per annum, and soil erosion has been cited as one of the greatest environmental problems facing Southern Africa. Refugees and internally displaced people in Mozambique, Malawi, Zimbabwe, Zambia and Ethiopia, who are forced by their desperate situation to seek shelter and energy, have removed trees. This has resulted in increased soil erosion, which produces an almost moonscape environment (Singh, 1996, 3).

Of course, there are also many Indian examples, such as those cases which have emerged in the the eastern states of Assam and Tripura. Bangladeshi immigrants, fleeing from famine and other environmentally-related occurrences, particular to their low-lying homelands, have "altered the local balances in land ownership, political power, ethnicity, and religion, stirring local resentments, riots, and an anti-immigrant movement" (Dabelko, 1999). The resultant violence claimed thousands of lives in the early 1980s and has flared up again in very recent times.

Water

Land degradation is closely associated with the critical issue of water, and water is an issue which crosses national and state borders with relative ease, reflecting the geophysical reality of rivers and waterways. Of all the Oceans, apart from being the poorest, the IOR posesses some of the driest areas on the planet, with Australia widely known by the title of 'the driest continent on earth'. In relation to the Middle East, Naff and Matson produced a relatively early work, entitled, *Water in the Middle East: Conflict or Cooperation?*. In 1984, they wrote as follows:

> Few regions of the planet offer a more varied physiography or richer mix of ethnicities (and) religions...Out of this compound, one issue emerges as the most conspicuous, cross-cutting and problematic: water - or rather its scarcity (Naff and Matson, 1984, 1).

Water was a major contributing factor in the Arab-Israeli conflict in 1967. During this period Israel was attempting to develop a nation-wide water supply and distribution system: the National Water Carrier (Shaheen, 2000, 139). The Arab League responded by making known its intentions to divert the river systems of the upper Jordan to Syria and Lebanon. Military attacks on Arab sites ended attempts at riperian diversion. Obviously, environmental security issues are not always the fundamental cause of wars. However, there is no doubt that, in this case, where there are long-standing ethnic and cultural divisions, a resource issue can be the 'straw that breaks the camel's back'.

In South Asia, there have already been intense disputes over the utilization of the Cauvery River between the states of Karnataka and Tamil Nadu. In 1991, several people were left dead and there was a temporary exodus of 100,000 Tamils from Karnataka.

In Australia, in the grip of a nation-wide drought 2002-3, the water issue, or lack of it, has become critical, straining relationships between several states with political boundaries traversed by its major river: the Murray. The World Health Organization released a report in 2000 stating that, unless profound changes are made to water management along the Murray, then cities like Adelaide, the capital of South Australia, which includes the Murray's mouth, will be unfit for habitation by the year 2050.

Obviously water issues include scarcity, as well as pollution, land degradation, and displacement concerns. Pollution of streams and water tables from waste from manufacturing industries; intensive agriculture; and that which emanates from imperfect engineering feats such as mega-dams and mass irrigation systems, is currently having dramatic effects on the quality of potable water sources. Many who live in the IOR live beyond the safety net of bottled 'mineral water', with women experiencing the brunt of carrying safe water long distances on a daily basis.

There is no doubt that the green revolution of the 1970s and 1980s increased crop productivity. However, its medium to long-term impacts are just being felt. For example, the system of Bangladeshi subterranean wells, hailed as a saviour of agriculture during the 1980s, is now contributing to the arsenic poisoning of millions of people, as the wells become leached by minerals released by rising water tables.

In India, the food bowl of the nation, the Punjab, is now in danger of over-irrigation, with three crops often being harvested per year since the green revolution. Pumps now run 24 hours a day on fully-subsidized electricity, and, for the first time, salinity is rearing its head as the water table rises to dangerous levels. One can only imagine South Asia's plight if the remarkably fertile Punjab became an ecological disaster zone.

Unfortunately, the future of water management in the poorer regions of the IOR has been vested in a faith in mega-technological solutions. Just when the more affluent world is lamenting the failure of mega-dams, the South is investing heavily in this outdated technology. The construction of these dams has led to the displacement of millions of people across the region. The most famous case, of course, is the Narmada project in Gujarat and Maharashtra. The mass displacement of peoples in the Narmada Valley has led to vital disconnections within and between communities and their place (Doyle, 2003).

In a rather sensationalist book, entitled *Water Wars*, Bulloch and Darwish write: "when next it (war) comes, as come it will, . . . every confrontation in the future will be affected by hydrography of the region. Water wars are on the way" (Bulloch and Darwish 1993).

Fisheries

The Indian Ocean contributes six million tonnes of marine fish production per annum. Most of the fishworkers live in coastal communities which are often considered risk communities due to "their isolation, fragile resource base and, often, poorer levels of education and health" (Chaturvedi, 1998, 713).

> The traditional and customary rights of coastal fishing communities have been eroded by the expansion of large-scale coastal tourism and industrial development...In their eagerness to secure valuable external earnings, states such as Indonesia and Thailand have frequently ignored the basic needs of local coastal communities and local women workers have subsequently migrated from, for instance,

Sri Lanka to the Maldives in search of work in the fish processing plants . . . (Chaturvedi, 1998, 713).

Aquaculture, a multi-national industry, is also changing the fabric of lives in these coastal communities, converting and destroying the ecological security of coastal farmlands and mangrove systems which sustain life, and replacing them with globalizing monocultures. The power of aquaculture industries is immense as they have been allowed to develop almost unchecked. Even in more affluent parts of the IOR, such as Australia, there are few regulations with teeth which effectively guide these industries. In some ways, this reflects the offshore nature of these activities which are largely away from the critical or regulatory eye. In a hallmark case of the Environment, Resource and Development Court (ERD) in 1999, the Conservation Council of South Australia (CCSA) challenged the State government, in conjunction with the powerful tuna fishing industry, over tuna feed-lots in Louth Bay on the Indian Ocean. Despite a successful outcome, with the CCSA proving that the industry did not adhere to the principles of environmentally sustainable development (ESD), the State government over-rode the court ruling, in favour of continued questionnable aquacultural practices.

Climate Change

Obviously, coastal, along with low-lying island communities, are also most at risk from climate change. Originally, climate change was considered as a classic environmental security issue in Northern terms. As aforesaid, the North came to comprehend that it had to share the same atmosphere and oceans with the South, and a continued policy of laissez-faire would be detrimental for the more affluent world. Also, climate change was a classic environmental issue portrayed as a natural 'force', rather than one which directly related to people. Consequently, climate change has not traditionally been seen as a key issue in the majority world. In recent times, however, it has assumed greater prominence amongst majority world environmentalists, due to the fact that some of the biggest state polluters and/or reliers on fossil fuels have not signed the climate change protocols in Kyoto and Johannesburg. Further tensions have emerged, due to the fact that both island states in the Indian Ocean, as well as poorer dwellers and coastal fishworkers on coastlines, will be the principal victims in global climate change. In this manner, climate change has metamorphosed from an elite, scientific, Northern issue into one which can usefully fit into the environmental justice agenda of the South.

At the end of October 2002, five thousand people from communities in India, including international NGOs, gathered in a Rally for Climate

Justice in New Delhi. This rally was organized to coincide with the United Nations meeting on climate change (Conference of Parties 8 - COP8), and was organized by the India Climate Justice Forum, including the National Alliance of People's Movements, the National Fishworkers Forum, Third World Network, and CorpWatch. In a press release, Friends of the Earth International (FOEI) wrote of the current frustration with climate change negotiations:

> Climate negotiations show no progress and communities are calling for urgent action to address climate change and to protect their livelihoods in a manner that is consistent with human rights, workers' rights, and environmental justice . . . Given the entrenched opposition to action from the fossil fuel industry and governments like the US and Saudi Arabia, environmental organizations joined forces with social movements in order to progress this most urgent agenda. The window of opportunity to prevent dangerous climate change is closing fast and, for many communities, the impacts are already alarmingly present (FOEI, 2002).

Nuclear Waste

Mention must be made of nuclear waste as an environmental security issue. Obviously, nuclear war is a major environmental disaster, and the nuclear fuel cycle, from uranium mining through to weapons testing, can have devastating impacts upon our environments. Australia at the moment is the one of the world's biggest uranium producers. It is also one of the proposed dumping grounds for the world's nuclear waste which is currently trying to locate spent fuel rods at a central repository. Pangea, a transnational waste company, has targetted Australia over recent years, most particularly the Indian Ocean rim states of South and Western Australia, as likely fuel dumps, due to geological and political security.

Obviously, tensions have increased in the region due to recent nuclear brinkmanship between India and Pakistan. However, let me limit most comments here to the potential for an accidental release from a nuclear reactor in the region. Jon Barnett and Steven Dovers, from Australia National University, explore this possibility within an environmental security framework in relation to Indonesia:

> According to the framework, the most difficult environmental problems are those with the most widespread impact (spatial scale), particularly if the impacts are spread across political boundaries. One study of the likely impacts of a release of radioactive material from Indonesia's proposed nuclear power programme shows that released

gases could travel as far as Australia, Brunei, Malaysia, Papua New Guinea, Singapore, Thailand, and, of course, through Indonesia itself (Barnett and Dovers, 2001, 164).

At Chernobyl, radioisotopes were spread over a 100,000 square kilometre area, and, "in terms of security, the magnitude of the clean-up after Chernobyl has been likened to the task of rebuilding after the German-Soviet war" (Barnett and Dovers, 2001, 164). With the higher levels of rainfall in the tropics of South East Asia, the leak would produce higher levels of radioactivity delivered by rainfall.

As with the case of mega-dams, in the poorer parts of the globe, such as the IOR, much faith and limited finances are being invested in technological solutions to environmental and resource issues; solutions which have now been effectively discarded by the more affluent minority world within their own national jurisdictions, whilst, at the same time, vociferously hawking their outmoded and dangerous wares in the markets of the South.

Population Movements/Environmental Refugees

Human overpopulation is an obvious and much quoted environmental security issue. The issue of environmentally-enforced population movements – environmental refugees – is of more relevance here due to its complex connectedness to all environmental issues I have previously touched upon in this Chapter. Deforestation contributes to climate change which exacerbates flooding in, for example, Bangladesh, which causes increased land degradation, leading to clean water shortages (impact of arsenic in wells), further exacerbating the siltation of rivers; decimating coastal fishing grounds; leading to enforced population movements into North-Eastern India, contributing to conflicts highlighting ethnic, religious and cultural differences. Kaplan imagines a not-too-distant-future as follows:

> . . . As the Maldive Islands, off the coast of India, sink into oblivion, and the shorelines of Egypt, Bangladesh, and South east Asia recede, tens of millions of people (are driven) inland where there is no room for them (Kaplan quoted in Curtis, 1998, 32).

The fact remains, however, that one does not have to imagine some hypothetical future scenario to realize the magnitude of this problem. The Indian Ocean is at once the home of the highest levels of refugee generation and possesses, by far, the top refugee-hosting states. In particular, Afghanistan, Rwanda, Liberia, Iraq, Somalia, Eritrea, Sudan,

Bangladesh and Burundi produce the bulk of the world's refugee movements. Interestingly, it is many of these same countries, with the inclusion of Iran, Pakistan, Tanzania and Ethiopia which are these refugees' new hosts. In the mid-1970s, these numbers added up to about 3 million refugees, and, by 1995, this number had escalated to 27 million, and this does not include those displaced within their own countries. It also does not include those displaced by development projects such as large dams within their own countries (Renner, 1996, 101).

Traditionally, refugees were defined in a manner which was appropriate to describe them after WWII: as people with 'a well-founded fear of persecution for reasons of race, religion, nationality, membership of a particular social group or political opinion' (Harrell-Bond quoted in Singh 1996). But now, the reality has forced a new definition, one that includes the concept of 'an environmental refugee'. Environmental refugees have now been defined as "people who fear that for environmental reasons they may not remain alive unless they migrate" (Zaba in Myers 199: 28).

As touched upon, internal movements within nation-states are often not recorded as environmental refugees, despite the fact that they often are even greater in number than transnational population movements. In the South African context, Singh states:

> People moving from rural areas in the Cape to the outskirts of Cape Town make up one of the fastest growing internally displaced groups in South Africa. Many of the displaced people were made redundant from white farms and were faced with the daunting choice of returning to impoverished homelands or trying to find a better life in the city. Since this phenomenon has only really started exploding in the 1990s, there is an absence of data and very few studies on the topic (Singh, 1996, 125-134).

Urban Explosion and Environmental Deterioration

Much of the stress on cities in states of the majority world of the IOR comes from the huge migration of environmental refugees from rural into urban areas. This migration, in 'western' terms is akin to the agrarian and industrial revolutions in its magnitude. These cities are increasing in population between 80,000 to 200,000 people per annum (Douglass et al., 1994). Of course, one of the additional causal factors of this migration is the globalization of trade, which has contributed to a decrease in the opportunities of small landholders and workers to make a subsistence living. Multi and transnational corporations can produce commodities far more 'efficiently', utilizing cheap labour markets, and the less stringent environmental demands of local legislation. Furthermore, urban environmental 'damage' is partly the result of local industries necessarily producing 'dirty' products in a bid to maintain competitiveness in the 'new global economy' (Doyle and McEachern, 1998, 79).

Furthermore, in many cities in the IOR (such as 'Bandung in Indonesia, Mumbai in India, or Bangkok in Thailand) where there is a massive population explosion, entire cities are forming outside of the established infrastructures of 'city limits' to accommodate these environmental refugees. In the urban context, environmental security issues include the provision of clean water; the physical labour of cleaning up refuse and the disposal of solid wastes; the building of shelter and the provision of sewerage systems; treating people directly for disease and malnutrition; direct provision of food and other basic essentials for living; and coordinating many other 'hands-on' tasks and activities (Doyle and McEachern, 1998, 94). Ultimately, environmental insecurity leads to poverty: poverty in terms of a lack of biodiversity, and, in terms of the incapacity of human beings to meet their most basic needs required for survival.

Conclusion

These seven brief synopses of some of the major environmental environmental security issues should not simply be read as a 'litany of woes'. All of these issues impact upon all nation-states in the Indian Ocean region. In traditional, or 'hard' security terms, environmental security issues, if not addressed, will lead to increases in human conflict and, ultimately, widescale disease, poverty and death. However, due to the regionally-shared nature of these problems, they are also issues which invite cooperation between nation-states; a shared agenda can emerge, with the potential for promoting a peaceful and extremely necessary dialogue. The long-term outcomes of such multilateral dialogues are immeasurable in positive terms.

The environmental security project of the Indian Ocean Research Group (IORG) is based on the assumption that all nation-states within the Indian Ocean region share a common interest in ascertaining and maintaining a secure environmental future. As part of this project, this Chapter aims to introduce a workable concept of a regionally-shared, basic and secure environmental future. Notions of environmental security issues within the region are multitudinous, differing greatly at times, from country to country and from sub-culture to sub-culture. Common appreciation of these diverse understandings and interpretations will be essential to ascertain during the initial stages of the research.

Research leading to environmental security must move through two further, major stages. The first stage is to collate base-line social and environmental data which identify the key environmental security issues within the region. This can be done by collating existing data, as well as pursuing field-work within different parts of the region. 'On-the-ground' research is essential in the context of environmental security issues, as the problems/issues must be valued and understood at the community level where most of the shared solutions lie.

Building on the data base, problem-solving strategies must then be identified and analysed. Ultimately the environmental security component of the IORG seeks to produce issue-identification and problems-solving techniques which can be used by a variety of community, non-governmental, governmental, and industrial partners in their implement-ation of solutions ascertaining and maintaining environmental security in the region. Ultimately, if solutions are to be pursued successfully, we must move away from the concept that environmental security uniquely concerns nation-states withstanding threats *from* the environment, and move to a position, which, in the words of Scrivener, "see environmental security as shifting the focus from state security to societal and individual well-being" (Scrivener, 2002, 184), advocating the concept of environmental security as security *for* the environment (of which humanity is a part).

Notes

1. Interestingly, since the events of September 11 2001, environmental security in the US context is often considered as one and the same as 'homeland security'. (Cheremisinoff 2002).
2. For an excellent example of this line of reasoning read Hartshorn, G.S., 'Key Environmental Issues for Developing Countries', *Journal of International Affairs*, vol. 7, 1991, pp. 393-401.

References

Africa News Service (2001) "US About-Face on Climate Change: A threat to Global Environmental Security", March 15.

Barnet, Jon and Dovers, Stephen, (2001) "Environmental Security, Sustainability and Policy", *Pacifica Review*, Volume 13, No. 2, June.

Barry, J. and Frankland, G. (2002) *International Encyclopedia of Environ-mental Politics,* Routledge, London and New York

Beck, U. (1995) *Ecological Enlightenment,* Humanities Press, New Jersey

Broda-Bahm, K. T. (1999) "Finding protection in definitions: the quest for environement security", *Argumentation and Advocacy*, Spring.

Bulloch, J. and Darwish, A. (1993) *Water wars: Coming Conflicts in the Middle East,* Victor Gollancz.

Chaturvedi, S. (1998) "Common security? Geopolitics, development, South Asia and the Indian Ocean", *Third World Quarterly*, Vol 19, No. 4, pp. 701-724

Cheremisinoff, N. P. (2002) "Environmental Security: The Need for International Policies", *Pollution Engineering,* May.

Curtis, J. (1998) "The Challenge of Environmental Security", *Habitat Australia*, Dec., vol. 26.

Dabelko, G.D. (1999), "The Environmental Factor", *The Wilson Quarterly*, Autumn, vol. 23.

Dabelko, G.D. and Dabelko, D.D. (1995) "Environmental Security: Issues of Conflict and Redefinition", Woodrow Wilson Environmental Change and Security Project Report, Issue 1: 3 – 12.

Department of Politics and Public Administration, University of Madras and Foundation for Sustainable Development (India) at the Indian Institute of Technology, Madras, Report, (2002) 'Citizen Action, Water Pollution and Public Health: An Analysis of Administrative and Implementation Dimensions', September.

Deudney, D. (1994) "Environmental Security", Book Review on *Ultimate Security: The Environemental Basis of Political Stability*, by Myers, N., W.W. Norton, New York, 1993

Dokken, K. (2001), "Environment, security and regionalism in the Asia-Pacific: is environemental security a useful concept?", *The Pacific Review,* Vol 14, No. 4, pp. 509-530.

Douglas, M., Y.S.F. Lee and K. Lowry, "Introduction to the Special Issue on Community Based Urban Environmental Management in Asia", *Asian Journal of Environmental Management,* 2, 1, ix-xv.

Doyle, T. (2000) *Green Power: The Environment Movement in Australia,* University of New South Wales Press, Sydney.

Doyle, T. and McEachern, D. (2001) *Environment and Politics, second edition,* Routledge, London and New York.

Doyle, T. (2003) "Dam Disputes in Australia and India: Appreciating Differences in Struggles for Sustainable Development", in Gopal, D. and Rumley, D., eds. *India and Australia: Issues and Opportunities* (New Delhi: Authors Press).

Foster, G. D. (2001) "Environmental Security: The Search for Strategic Legitimacy", *Armed Forces and Society,* Spring, vol. 27, 13.

Friends of the Earth (2002), Press Release, Tuesday 29 October, "Urgent Action needed to address climate change".

Kakonen, J. (1994) *Green Security or Militarized Environment*, Dartmouth Publishing Company, Aldershot and Brookfield.

Hassan, S. (1991), *Environmental Issues and Security in South Asia* (London: Brassey's for the IISS).

Hugo, G. (1996) "Environmental Concerns and International Migration", *International Migration Review,* Centre For Migration Studies, vol. 30, Spring.

Imura, H. (1994) "Japan's Environmental Balancing Act", *Asian Survey,* XXXIV, 4, pp. 355-368.

International Monetary Fund cited in R. Broad (1994) "The Poor and the Environment: Friends or Foes?", in *World Development*, 22, 6, pp. 811-822.

Levy, M. A. (1999) "Exploring environment-security connections" (World Wide Web sites on the relaitonship between environment and security), *Environment,* Jan, vol. 41.

Lowi, Miriam R. and Shaw, Brian R. (2000) *Environment and Security Discourses and Practices,* Macmillan Press Ltd, London and New York

Lukong, P.F. (2002) "Dealing with HIV/AIDS in Sub Saharan Africa: Benefits and Challenges of a Spatial Approach in Cameroon", Presented at the Joint AURISA and Institution of Surveyors Conferenc, Adelaide, South Australia, 25-30 November 2002

Madsen, Stig Toft (1999) *State, Society and the Environment in South Asia*, Curzon, Surrey

Moench, Marcus, Caspari, Elisabeth and Dixit, Ajaya (1999) *Rethinking the Mosaic: Investigations into Local Water Management*, Nepal Water Conservation Foundation, Kathmandu and the Institute for Social and Environmental Transition, Boulder, Colorado, U.S.A

Moyo, P., O'Keefe and N. Middleton, (1993) *The Tears of the Crocodile: From Rio to Reality in the Developing World,* London: Pluto Press: 5.

Myers, N. (1995) *Environmental Exodus. An Emergent Crises in the Global Arena,* Climate Institute: United States.

Naff, T. and Matson, R. (1984) *Water in the Middle East: Conflict or Cooperation,* Westview Press, Boulder.

Pettiford, Lloyd and Curley, Melissa (1999) *Changing Security Agendas and the Third World,* Pinter, London and New York

Redclift, M. (2000) "Addressing the Causes of Conflict: Human Security and Environmental Responsibilities", *Reciel,* 9 (1).

Schultz, Richard H., Godson, Roy and Quester, George (1997) *Security Studies for the 21st Century,* Brasseys, Virginia

Shaheen, M. (2000) "Questioning the Water-War Phenomenon in the Jordan Basin", *Middle East Policy,* Vol VII, No. 3, June.

Singh, M. (1996) "Environmental security and displaced people in Southern Africa", *Social Justice,* Winter, vol. 23, no. 4, pp. 125-9

Soroos, M.S. (1995) "Environmental Security: Choices for the twenty-first century", *National Forum,* Winter, vol. 75, no. 1, pp. 20-5

Thompson, M. (1997) "Security and Solidarity: and anti-reductionist framework for thinking about the relationship between us and the rest of nature", *The Geographical Journal,* Vol. 163, No. 2, July, pp. 141-149.

CHAPTER 9

Re-envisioning Transboundary Water Disputes as Developmental Conflicts

Radha D'Souza

Conflicts over transboundary watercourses are seen as important sources of geopolitical conflicts in the world today. Typically, transboundary water conflicts are conceptualized as political conflicts between states over water sharing. This Chapter argues that such a conception is often too simplistic and that water conflicts are another form of development conflicts. 'Development' as a global project is usually critiqued within the framework of citizen-state tensions and, sometimes, for the impact it has on questions of community, class, caste, race, and ethnic tensions. This Chapter argues that the international conflicts between states are equally underpinned by the ways in which 'development' is conceptualized by international development agencies and within international organizations under the UN system, particularly the Bretton Woods' organizations. The 'development' process has resulted in unequal regional development that manifests as 'North'-'South' tensions, on the one hand, and interstate geopolitical conflicts, on the other. This Chapter discusses the policies and programs promoted by the UN organ, Economic and Social Council (ECOSOC) and the Economic and Social Commission for Asia and the Pacific (ESCAP) under it, with regard to the regulation of rivers, and argues that the way in which river basin development and regulation of rivers is conceptualized by the UN organs is the principal source of geopolitical conflicts between states. Such a 're-envisioning' is essential if we are to understand and thus resolve such conflicts in the Indian Ocean Region.

Introduction

The trajectory of the discourse on dams and development, spanning the entire course of the post-war world order, has undergone a major about face during the past decade or so. From being the central issue in development, it has now become the most contentious issue in development (McCully, 1996; The World Conservation Union and The World Bank, 1997). Conventional development theory in the post-war era, with its focus on 'forced take-off', econometric modelling, development planning, aid and development assistance, promised elimination of poverty and standards of living comparable to the 'West' for the 'developing' states (Meier, Seers and Bauer, 1984; Meier and Schultz, 1987). The case for large dams was premised on those theories and informed international organizations, bilateral and multilateral aid and UN institutions.[1] The failure of 'development' to deliver on its promises led to a wide-ranging critique of development theory (Kothari, 1988; Leys, 1996; Moore and Schmitz, 1995). The critique of dams and development in 'developing' countries draws inspiration from the wider critique of development theory. The critique focuses on the impact of large dams on communities, class, capital, gender, ethnicity, poverty and social justice (Fisher, 1995; Goldman, 1994; Thukral, 1992). Environmental movements in the 'West' add another strand to the critique of dams and development (Arnold and Guha, 1995; Guha, 1997; McCully, 1996; Singh, 1997; Worster, 1983). Movements such as the International Rivers Network have helped develop an awareness of the extent, scope and universality of the experiences in relation to dams and development. Despite such an extensive critique, the ramifications of the social impact of dams and development for transboundary tensions in international relations are less understood.

The critique of dams and development does not, generally, touch upon transboundary issues in dams and development. At best it provides a descriptive account of disputes between states that touch on apparent causes, not sociological ones (Crow, 1995; Crow and Singh, 2000; Tvedt, 1986). Nevertheless, if dams cause so much grief *within* societies and have such deleterious effects on disadvantaged sections *within* states, surely they must have consequences for international relations to the extent that domestic pressures within nation-states do influence the international relations policies of states. This omission may be attributed to the way the 'national' and the 'international' are conceptualised in social theory. Relying primarily on juridical attributes of the state as the primary unit of analysis, the national and international dimensions of development conflicts are viewed as operating in distinct *realms* that manifest discursively as the internal-external aspects of development. Thus in the international *realm*, developmental conflicts are envisaged as conflicts between the countries of the 'North' and the 'South'. In the national *realm*, developmental conflicts are envisaged as conflicts among citizens, or between citizens and the state.

Internally, within society, the nation-state has long remained the subject of extensive historical and sociological critique. Benedict Anderson's (1983) work pointed to the colonial construction of nation-states in the 'developing' countries and the complexities of nationalism in societies with colonial histories. Amy Chua (1995) highlights the wider international macroeconomic policy implications for nationalization and privatization and their impact upon specific communities and societies in definite ways in 'developing' countries. The political economy of ethnic and political strife in 'developing' states is the subject of a growing body of literature on political and regional studies. These studies touch on the impact of political economy on ethnic, religious, racial and communal tensions and indirectly allude to the ways in which international economic policy impinges upon citizen-state relations. They also direct attention to the colonial inheritances of the nation-state in the 'developing' countries. However, they do not generally draw attention to international macroeconomic policies as a factor in international relations nor to the possibility that development conflicts may manifest as interstate conflicts.

Externally, in international relations, sociological categories of analysis such as class, capital, gender, ethnicity, race, and others are used more as descriptive terms than analytical categories. Nation-states remain the basic unit of analysis in international relations theories, although the pressures on the nation-state due to economic, social, political and environmental reasons, are widely acknowledged (Arrighi, 1993; Cox, 1993 a., 1993 b.; Gill, 1991, 1993; Halliday, 1992). The critique of 'globalization' problematizes the nation-state, focusing largely on a critique of conventional liberal theories of the state as a point of departure, and, taking positions along the state-market duality as the two ends of a continuum.

The conceptual and analytical tools used to explain the tensions in the two *realms* are inadequate to explain transboundary conflicts that arise from the developmental process. In the case of dams and development, it does not provide us with adequate critical conceptual tools to make the connections between developmental conflicts at the international and national levels.

Consequently, transboundary conflicts over water in the political literature tend to be analysed as geopolitical issues between nation-states arising from an undifferentiated and homogenous 'national interest', on the one hand. The geopolitical lens through which transboundary water conflicts are envisioned has spawned a growing literature on 'water wars' as an important global security risk (Anan, 2002; Gleick, 1993; Hoffbuhr,

2002; Starr, 1991). On the other hand, the anti-globalization literature has sharpened the critique of development by highlighting corporate interests in water projects and the ways in which such interests intertwine with the goals and purposes of international development organizations (Bakker, 1999; Barlow and Clarke, 2002; Shiva, 2002). However, the connections between the problematic 'development' agenda and transboundary conflicts over water are tenuous at best.

This Chapter attempts to develop a framework of analysis for making the connections between political conflicts arising out of transboundary water sharing issues and the international dynamic of 'development' that underpins those conflicts. It argues that transboundary water conflicts are developmental conflicts at the international level that need to be understood within the broader dynamic of post-war imperialism.

Water Conflicts: A Stratified View

Conflicts over water may be analysed at several levels – micro-sociological, meso-sociological and meta-sociological. At the micro-sociological level, conflicts over water involve conflicts *among* citizens. In the case of canal irrigation they could involve conflicts between "head-enders" and "tail-enders" (Folke, 1998). Water conflicts could manifest as conflicts between castes, communities, religious, and linguistic groups, who, due to their geographic location in relation to irrigation projects, become winners and losers. Before project construction begins, conflicts over water involve conflicts between citizens and the state on rehabilitation, resettlement, price of land, submergence of land and compensation issues (Fisher, 1995; Thukral, 1992). These issues are well traversed by the literature on dams and development. However, the issues at the micro-sociological level arise because of policies, programmes and plans that are undertaken at the meso-sociological level.

The meso-sociological level may be seen at the level of national policies and programmes on water. At the meso-sociological level, conflicts over water have an internal dimension *and* an external dimension. At that level, water conflicts arise because of federal/national planning, state policies, government programmes where different regions and interests vie for projects, state finances, subsidies, pricing policies and concessions and legal and regulatory mechanisms. At the meso-sociological level, the winners and losers in development planning for water resources results in unequal development among regions and/or states and manifests as the politics of regionalism. At the meso-sociological level, conflicts over water also include an international dimension. They entail tensions with international development agencies, bilateral and multilateral aid policies, repayment schedules, supervision of projects and tensions between domestic policies to meet the expectations of citizen groups and policies to fulfil loan and finance criteria by international lending institutions and governments.

The meso-sociological level becomes the generative structure for conflicts at the micro-sociological level. How the tensions are negotiated and managed at the meso-sociological level influences and impacts upon how water disputes are perceived by citizens and the kind of water conflicts that arise at the micro-sociological level. Meso-sociological discourses on water resources create the lens through which water issues come to be perceived at the micro-sociological level. Whether water conflicts flare up into major federal-state conflicts, or communal strife, or, whether they are seen through a nationalist lens as problems created by international development agencies; or, whether they are seen as issues caused by the ideology and politics of political regimes; or, a varying combination of all of the above, depends to a considerable extent on the meso-sociological discourses on water resources.

There is a third level at which water conflicts need to be analysed that relates to the theme of this Chapter. Water conflicts may be analysed at the meta-sociological level. At that level, water conflicts may be located within capitalism and imperialism as the meta-sociological structure for society today. Nation-states, international territorial boundaries, the UN as the institutional umbrella of the post-world war order, 'development' and 'progress' as global projects, are all inextricably intertwined with capitalism and imperialism. However, envisioning 'the national' and 'the international' as distinct realms, and the embedded nature of what Peter Taylor (Taylor 1994) calls a 'container view of the state', fetters the way transboundary water conflicts are envisioned. The way transboundary water conflicts are envisioned at present fetishses the *real* nature of the conflicts as conflicts having their source in the generative structures of contemporary capitalism and imperialism.

Susan Buck-Morss (Buck-Morss 1995) argues that it is the depersonalization of exchange within capitalist society that depoliticizes economic power, no matter how close capitalists and politicians may become. Liberal democratic traditions exist not on the political notion of nationalism, but on the economic notion of a collective based on the depersonalized exchange of goods. The economy is founded, not on any empirical reality - for example, place, community, history and so on – but on a perceptual apparatus based on representational mapping that objectifies it. Visualization of the economy is premised on statistical representations, quantitative reasoning and mechanical causation. By analogy, the post-war world order universalized liberal traditions by institutionalizing them in law and institutions under the UN system. The

UN framework is founded on conventional liberal conceptions of economy and society. Within the liberal UN framework, nationalism and nation-states were envisaged as a tier in the hierarchy of post-war institutions that would engage in the denationalized exchanges in the post-war world and thereby 'depoliticize' economic power in the New World Order that emerged at the end of the second world war. The legal and institutional framework in the UN system was designed to depoliticize economic power, no matter how close capitalist states and international organizations were to become. De-fetishizing the meta-sociological structures of post-war capitalism and imperialism as generative structures for transboundary water conflicts today must therefore begin by interrogating the role of international institutions and international law in the regime of post-war capitalism and imperialism.

Development as a Post War 'Project'

The World Wars saw the breakdown of a wide range of institutions and infrastructures on which capitalism and imperialism were based. Relations between colonies and capitalist states were one such relationship (D'Souza, 2002 d.). In the post-war world, reconstruction of the relations between the colonies and the imperial centres were premised on two distinct types of discourses. The first was a discourse about nationalism, democracy and a constitutional order modelled after western liberalism. The second was a discourse about economic 'development', also modelled on assumptions about replicating the experiences of "Western" capitalism in the rest of the world. The discourse of nationalism entailed devolution of political power to domestic governments and local authorities. The discourse of development entailed conceding economic power to international organizations and centres of capital. The apparently opposing movements in politics and economics are held together by the legal and institutional framework that evolved during the colonial era, especially the era just preceding national independence (D'Souza, 2001 b.).

Imperialism under the Empire system had cobbled together assortments of nations, tribes, peoples into nation-states that had trappings of liberal 'Western' states, but without the historical and geographical foundations. The inter-imperialist rivalries of the colonial-empire era were influential in the ways in which the boundaries of the colonial states came to be drawn. The colonial 'project' entailed development of legal and institutional structures within the colonies for reproduction of colonial-imperial ties (D'Souza, 2002 c., 2002 d.; Washbrook, 1997). The politics of democracy and "responsible government", as it came to be called in the British Empire, entailed appropriate law and institutions that facilitated the devolution of power within a stratified power structure. It created the conditions for national independence to occur and facilitated the reconstruction of the colonial-imperial ties ruptured during the World Wars.

National independence was founded on conceptions of the economy that were premised on 'modernization' and a vision of capitalist development without the colonies. In many colonies, including India, conceptions of 'development' and new divisions of labour arose during the inter-war years when the colonies became suppliers of resources, human and natural, for the war. In the mature colonies, including India, industrialization strategies have their genesis in wartime necessities. They included power generation and multipurpose river basin development projects which became an important feature of post-war development planning internationally within the UN organs and development agencies that were formed at the end of the World Wars.

Philip McMichael (1996) describes 'development' as a global 'project' based on an international division of labour that became the basis for global commodity chains to emerge in the post-war era. He shows how the development 'project' had its genesis in the colonial divisions of labour and the commodity chains within the Empires of those times. Global divisions of labour and commodity chains, however, require a legal and institutional framework to grow and to function as recurring everyday practices. Global divisions of labour also require global infrastructure development on which the commodity chains and divisions of labour could be founded. Cheap labour and cheap natural resources from the colonies were an intrinsic feature of the colonial division of labour, and remain an intrinsic feature of post-war divisions of labour. However, the legal and institutional regimes on which they are founded in the post-war world are distinctive.

The development 'project' in the post-war era was an ideological, economic, legal, political, cultural and scientific/technological one premised on definite ways in which nature-society-human relationships came to be envisioned and institutionalized in the structure of the United Nations. It facilitated mobilization of resources, material and ideological, for post-war capitalism and imperialism. The United Nations was thus the institutional umbrella for the development 'project' and for mobilization of resources for post-war capitalism and imperialism.

The Development 'Project' Under the UN Umbrella

The United Nations arose out of the ashes of the League of Nations, which collapsed under the pressure of inter-imperialist rivalries and the difficulties the United States, the rising imperial power then, had with the structure of the League. An important tension was the strict and often legalistic conceptions of state sovereignty on which the constitution of the League was founded and the demands of an expansionist monopoly capitalism, of which the US was the centre, at that time. The tension is captured by the rapid expansion of international organizations, public and private, in wide-ranging areas from science and technology to 'development' in the colonies, at a time when, politically, relations among the imperial powers were at their worst (Bowett, 1964). Prior to World War I, it is estimated that there were at least 400 public and private international organizations (Chimni, 1993). The League was restrained in advancing the public international organizations as it was founded on a membership of states zealous about safeguarding the sovereignty principle. Equally, it was restrained from nurturing the private organizations seeking standardization of infrastructure and investments, from postal unions and industry organizations, from dam building to road laying, due to the inflexible ways in the which the public and private lines were drawn. Yet, infrastructure was a prerequisite for capitalist expansion. In the absence of an appropriate institutional and legal framework for monopoly capital to emerge and expand, there could not be "level playing fields", market institutions and infrastructures for monopoly capitalism. At the turn of the 20th century, capitalism had outgrown the three institutions on which it was founded – the nation-states, the empire system and international law and institutions. The United Nations addressed all three institutions.

The structure and the character of the United Nations was conceptualized largely by Britain, the declining imperial power, and the United States, the ascending imperial power, between 1938 to 1942, before the Dumburton Oak document was signed bringing in the rest of the world. The bilateral agreements between the United States and Great Britain during that period facilitated the transition of the leadership of capitalism and imperialism from Britain to the US, which, in turn, allowed the Allies to have a dominant role in the UN. The structure of the United Nations brings together economic and political organs to create a legal regime and institutional order for the post-war political economy in the interests of post-war capitalism and imperialism. Politically, the members of the United Nations are nation-states. Responsibility for political, security, and military matters is assigned to the Security Council, where the veto power was granted permanently to five countries, three of them, Allies in the World War, ensure that the leadership of the United Nations remains with the Allies. Thus, it safeguards the state sovereignty principle of the earlier era while assuring that political leadership remains in the hands of the victors in the World War, the Allies.

The United Nations Charter envisaged an economic role for the United Nations that the League did not have. Chapter IX and X sections 55 to 72 of the Charter clearly prescribes a central role for the United Nations in economic matters. The status of the Specialized Agencies within the United Nations creates the institutions through which it fulfils this role. The Specialized Agencies have their own constitutions and special relationship agreements with the United Nations. Their constitutions and relationship agreements do several things. First they define the extent to which the political and social organs of the United Nations can interfere in the economic organs (Brown 1992). Second, they allow for the membership of private international organizations, including science and technology organizations, industry organizations and corporations to become involved in United Nations' programmes and policies. Third, their constitutions are drawn up in ways that ensure that the dominant capitalist powers remain in control of the economic regime. Fourth, they ensure the production of knowledge and discourses necessary for the post-war regime. Fifth, they are structured to facilitate the selective involvement, in certain specialized agencies, of states, disempowered within the Security Council, to participate in the programmes and policies of the United Nations. Sixth, the Specialized Agencies blur traditional divisions of public and private by bringing together states and private sector organizations in common programmes, thereby limiting the options of states on economic issues, leaving them to fit into a hierarchy within a global power structure. The Specialized Agencies are of two kinds. The first, the Bretton Woods organizations, regulate the financial regime for post-war capitalism by creating the institutional framework for banking, finance and trade. Their autonomy within the United Nations is safeguarded in their specialized agreements with the United Nations. Their constitutions also safeguard the leading roles for the dominant capitalist powers (Brown, 1992). The other Specialized Agencies (for example, UNESCO, WMO, FAO) create the infrastructure required for post-war capitalism and imperialism. The Agencies are more inclusive in that they bring together public and private interests, states, developed and developing, and other 'civil society' representatives.

From its inception, the law-making role assigned to the United Nations in the Charter was far more extensive than that of the League. Codification and development of international law is enshrined in the charter as a central role for the United Nations. The law that UN organs must codify is the 'law of civilized nations'. At the time of the formation of the United Nations, Euro-American legal systems had displaced indigenous law in most parts of the world due to colonization. The 'law of civilized nations' therefore means Euro-American law drawn from Roman law, cannon law and the merchant law traditions of Europe within the institutional schema of public and private domains in society mediated by the state. The role in the Charter for development and codification of the 'law of civilized nations' was in conjunction with the older International Court of Justice under the League, which was limited to resolving inter-state matters with the consent of the states. The new role for law was much wider. The International Law Commission has, over the past fifty-five years, codified international law on specific issues from the law of shipping to transbounday watercourses. Other international legal forums have emerged, including the international war crimes tribunal and criminal justice as a result of the UN's role in developing the law.

Finally, the Economic and Social Council of the UN (ECOSOC), an important organ of the United Nations, brings together all three of the above to play a coordinating role. The ECOSOC is a new invention in the UN as there was nothing like it in the League, and this was one of its main limitations, noted above. The conceptualization of ECOSOC underwent the most extensive changes during the formative period when the UN was conceptualized. It is the forum where social and economic agendas are introduced. It brings together the Specialized Agencies, the states and the non-governmental organizations around key themes and issues. It disseminates the decisions of ECOSOC through the regional organizations, for example, Economic and Social Commission for Asia and the Pacific (ESCAP), Economic Commission for Africa (ECA), Economic Commission for Latin America and the Caribbean (ECLAC) and Economic and Social Commission for Western Asia (ECWA). The regional forums of the ECOSOC provide a platform for ironing out the differences between the developed and developing states, on the one hand, and public and private organizations, on the other. It is not accidental that the dependency theories of the 1950s and 1960s were first articulated within the ECLAC and found a voice within the ECOSOC's regional organizations. From its inception, the 'development' of the newly-independent colonies was a central preoccupation of ECOSOC. From its inception, decolonization and development went hand in hand in the New World Order under the UN in the post-war era.

'Development' was introduced into the ECOSOC soon after President Truman of the US promised in January 1949 and outlined what came to be called the Four Point Programme to the US House Committee on Foreign Affairs, which took legislative shape in the form of the International Technical Co-operation Act of 1949. The Act envisaged external assistance programmes of two types: first, the technical, scientific and managerial knowledge necessary for 'development', and, second, the machinery, equipment and expansion of productive capacity. The legislation envisaged technical cooperation as a multilateral enterprise promoted through the United Nations (Khan, 1961).

River Basin 'Development' And The UN Family Of Institutions

River basin development was an important component of the wider development 'project'. The development 'project' that McMichael (1996) describes was founded on cheap labour and an industrial infrastructure, of which cheap food and cheap hydropower were necessary components. One of the first resolutions brought to ECOSOC was by the United States and it was on 'development' in 1948. Flood control and development of arid regions were amongst the first issues that came to be considered under ECOSOC's 'development' plans for the newly-independent states that were still emerging at that time (United Nations, 1948-49, 1950, 1951, 1952). The initial resolutions on 'development' generally, and river basin development in particular, helped mobilize resources, economic, political, scientific and technological, intellectual, cultural and ethical for river basin development.

River basin development was conceptualized as a series of projects on rivers. The project orientation isolated the economic dimensions of river basin development and insulated it from the social, environmental, cultural, class and regional implications. Conceptualized as a series of projects, river basin development could be discussed in the language of investments and returns, development planning and technical assistance, aid and development lending without regard to the society in which the projects must necessarily exist. The project approach enabled the waters of a river to be packaged for investment to international lending agencies. Thus, river basin projects in developing countries have a very different trajectory of historical development from the 'Western' states.

Nearly all UN agencies, financial, scientific and technological and others, promoted the idea of large multipurpose river basin projects as part of the development agenda. Following the resolution on development, flood control and development of arid regions, the ECOSOC brought together the Specialized Agencies, financial and technical, various private and public international organizations, governments, bilateral and multilateral development agencies from the 'developed' and 'underdeveloped' countries on a range of issues necessary for river basin development. They included standardizing procedures for data collection, mapping water resources, including rivers, groundwater, surface water, design and construction of projects, promoting science, technology, education and manpower planning as part of development planning, vocational training and promoting hydrology as a discipline. The regional organizations of ECOSOC emerged as the site where ideas of river basin development took form and shape and translated into policies that were compatible within the nation-states of the region, the international organizations and developed capitalist states (D'Souza, 2002 c., 2002 d.).

The First Regional Conference of the regional organization for Asia, Economic and Social Commission for Asia and the Far East (ESCAFE), later reconstituted as the Economic Commission for Asia and the Pacific (ESCAP), in 1951, was on flood control. The Conference resolved that "flood control could not be divorced from unified river basin development". Subsequent regional programmes brought together a number of government and private agencies from the 'developed' and 'underdeveloped' states for projects. If cartography and mapping of land and minerals were essential to the colonial project, in the post-war era, mapping turned to water resources first and later to other biological resources. The 14th, 16th and 17th sessions of the ESCAFE promoted ground water studies in collaboration with UNESCO, the US Bureau of Technical Assistance, France, USSR, USA, WHO, WMO, the International Commission on Irrigation and Drainage (ICID) and private consultants from the Netherlands and the United States. Standardization of water resource planning in the countries of the regions emerged as an important concern (Economic Commission For Asia And The Far East, 1965). ECOSOC undertook a survey of major international rivers and published *A Compendium of Major International Rivers* in collaboration with UNESCO that became an important source of data for the international dam construction industry (Economic Commission For Asia And The Far East, 1966). From the mid-1960s, hydrological data became an important part of the regional agenda of ESCAFE. The purpose of the data was clear:

With the advent of financial assistance to developing countries on an extensive scale, recognition has extended to economists and banking authorities whose insistence on the provision of adequate data, properly analysed and interpreted, has been a cornerstone of development policy (Economic Commission For Asia And The Far East 1966, 3).

Hydrological data were an important requirement for standardization of equipment and procedure. However, the inadequacy of data was not to hold up dam construction projects. In 1962, the Fifth Regional Technical Conference of the ESCAFE reported:

The Conference realized the importance of hydrological data but felt that, to avoid delaying essential and urgently needed projects on account of insufficiency of such data, resort should be made to indirect hydrologic determination accompanied by realization of the calculated risk entailed. It was pointed out that, in the case of integrated development, it might not be possible to undertake all of the projects in a comprehensive master plan simultaneously. Hence careful consideration of priorities was required in the selection of projects for early construction and... notwithstanding the deficiencies in flood and other hydrological data, the planning of flood control works based on existing data should proceed and improvements be carried out as and when more data become available (Economic Commission For Asia And The Far East 1965, 3-4).

ESCAFE's initiatives rationalized dam construction notwithstanding the absence of water resources data and social and environmental costs and gave it a legitimacy that private industry organizations could never have succeeded in doing. The guidelines for water resource planning emphasized standardizing procedures and criteria for economic and financial analysis of water resources planning, general measurements of standards, identification of measurement of costs, identification and measurement of economic project benefits, criteria of economic evaluation and financial evaluation and observed:

> . . . because development can seldom be financed out of revenue, it mortgages the future (Economic Commission For Asia And The Far East ,1968, 81).

The ESCAFE made a curious distinction between 'financial analysis' and 'economic analysis' in the guidelines for water resource planning by states. Financial analysis developed criteria to analyse the project costs and returns on investments during the maturity of the loans raised to finance it. The 'economic analysis' involved criteria that took into account all benefits, tangible and intangible, primary and secondary, and all costs induced and associated costs as well as installation operation and maintenance costs. The financial analysis included only the direct monetary returns or revenues (Economic Commission For Asia And The Far East, 1968, 81). This distinction enabled the international bilateral and multilateral institutions and public and private organizations to lend money without any legal risks for the social and environmental consequences. In the words of ESCAFE:

> Irrigation projects, which are seldom in the position of showing a profit, though capable of meeting foreign exchange costs and costs of operation and maintenance may end up showing a loss in local currency at the end of the period of analysis. The magnitude of the costs is a critical factor in determining project viability. If, at the end of the period of analysis the anticipated revenue is sufficient to meet

interest charges but little more after paying operation and maintenance expenses, the unpaid principal may be regarded as perpetual debt. It is, however, in the national interest to repay the outstanding principal, and persistent efforts should be made to keep water and other rates high enough to do so (Economic Commission For Asia And The Far East, 1968, 83).

On the other hand, with regard to project costs, cost structure and amortization of costs, the planning guidelines required of states that:

> . . . the annual sum to be set aside to amortize a loan is a function of the interest rate, grace period and maturity. These terms are dictated by the lending institutions and, in the case of foreign loans, are quite rigid. Local government loans may be flexible, especially regarding the repayment period and the rate of repayment (Economic Commission For Asia And The Far East, 1968, 84).

The distinction enabled the international agencies to lend based on 'financial analysis' and the national governments to rationalize expenditure on dams based on 'economic analysis'.

Thus, the UN, through its regional organs and agencies, did two things. First, it insulated the international development agencies by negotiating consent from the states through the activities of its regional organs; second, it devised the legal and institutional means for offloading the social and environmental responsibilities following from river basin development to the nation-states.

The World Bank and other regional development banks under its umbrella, such as the Asian Development Bank, mobilized resources internationally to support construction of large scale river basin projects as part of 'development' finance (Imhof, 1997; Sklar and McCully, 1994). UNESCO and WMO, amongst others, mobilized the educational, scientific and technological institutions for large-scale multipurpose dams based on the idea that technology was the answer to problems arising from colonization. A number of private international organizations also expanded in the post-war era from their Euro-American focus to many developing states. Most prominent amongst the private international organizations was the International Commission on Large Dams (ICLD), the premier international organization of the dam industry.

Within the UN, the launch of special decades, conferences and conventions helped bring together a wide range of agencies and states, public and private, around key themes and issues (Taylor, 1989; Willetts, 1989). The Development Decades, first (1961-70), second (1971-80) and third (1981-90), coordinated a wide range of development agencies and organizations. The United Nations Development Programme (UNDP) institutionalized the co-ordination and rationalized the financing of development internationally. The International Hydrological Decade, launched in 1965, helped to co-ordinate the scientific and technological agencies involved in river basin development. The International Hydrological Programme (IHP) institutionalized the co-ordination and standardized the technological and infrastructure requirements for large-scale multipurpose projects to expand internationally.

The dam industry in 'Western' capitalist states evolved in response to large-scale industries and the demand for hydropower during the inter-war years (McCully, 1996; Teclaff, 1967; Teclaff, 1991). In the post-war era, the dam industry was ready to expand internationally. The mobilization of resources by the UN and its agencies created the institutional conditions required for the international expansion of the dam industry.. It facilitated mobilizing from the 'top' financial, technological and ideological resources for the post-war expansion of capitalism (D'Souza, 2002 c., 2002 d.).

This broad and necessarily brushstroke overview of the UN as the structural umbrella for the development 'project' in the post-war era is important for a number of reasons. First, the critique of dams and development draws attention to the World Bank and other regional development banks for their lending policies for large dams. The critique is limiting, as it does not contextualize the World Bank and the development 'project' in the post-war era. Second, the critique employs two main analytical tools the 'poor' in the 'Third World', on the one hand, and global capital, represented by agencies such as the World Bank, on the other. The state, both in the 'First' and 'Third' worlds, and their place between the global rich and poor, remains primarily a political one in the conventional sense. Economic and politics once again appear to operate in different realms. Third, envisioning dams and development as a matter between the poor in the 'Third World' and global capitalism symbolized by the international development finance agencies, such as the World Bank, fetishizes the institutional context for development lending and the UN as the structural umbrella under which they must exist. Consequently, it leaves the door open for the politics of water to conjure spectres of water wars as a tool in geopolitical imagination, and transboundary tensions over water continue to be articulated in the more conventional language of geopolitics. The conceptualization leaves open a conceptual disjuncture between dams and development viewed as economic issues and transboundary tensions on water viewed as political issues. The gap echoes the institutional separation of 'development' as a comprehensive 'project' of post-war capitalism and imperialism, and nationalism as an integral part of the structure of the post-war world order.

Nationalism and the Development 'Project'

One major conceptual problem in understanding questions about 'developing' societies, or ex-colonial societies, is explaining the persistence and continuities of colonial features of society much after independence (Alavi, 1975; Harshe, 1997; Washbrook, 1997). Whereas the institutional features of imperialism in the colonial era are more readily understood, the institutional features of post-war imperialism are mystified due to the dual role of the nation-state in ex-colonial societies. Politically, independent nation-states fulfil people's aspirations for self-determination. Economically, nation-states in ex-colonial societies were formed and shaped by colonialism. At the time of formal political independence, they were already integrated into the economic structures that emerged in the post-war era. In many colonies, the inter-war years were periods of major constitutional and institutional changes that dovetailed the changes and the transformations within imperialist states (D'Souza, 2001 a., 2001 b.). In the post-war world order, then, imperialism came to be envisioned primarily as an economic phenomenon. The political and institutional aspects were not subjected to the same level of critical interrogation as the economic aspects.

The states in developing countries face a structural conundrum. Economically, the states find themselves victims of development, with debt-burdens, non-performing water projects, and environmental stress. Given the inheritance of colonial boundaries, the multiethnic composition of states and the superimposition of colonial political boundaries over watercourses, the social conflicts from river basin development manifest as ethnic, religious or communal tensions that have the potential to tear the state apart. Disempowered within the international economic institutions and meta-level power structures, and confronted with civil strife following from the development 'project' at the micro-level, one possible way for the state in ex-colonial societies to save national unity *and at the same time* stay within the economic parameters of the post-war institutions, is to deflect the crisis to another neighbouring state at the meso-level. If there is no water in the rivers, or irrigation systems fail, the problem is not with the flawed development strategy but with the neighbouring state.

In the post-war world, the discourse of nationalism emerged as a political and cultural discourse. The politics of post-coloniality ensured that the problems of 'developing' societies were envisioned as cultural and social problems that were colonial inheritances (Dirlik, 1994). The cultural lens through which post-colonialism was envisioned delinked past imperialism from the present. The discourse of 'development' was premised on the assumption that the state in ex-colonial societies could, through sheer 'will', bring about capitalism without the colonies. This economic and political conceptualization envisioned the state in ways that limited the deconstruction of the legal and institutional features of post-war imperialism. More importantly, it mystified the role of the UN as the institutional umbrella for post-war capitalism and imperialism. Envisioning the problems of the post-war world within the framework of nation-states meant that the explanatory categories available for understanding problems of transboundary water conflicts were contained within nation-states. 'Development' as the underlying cause of interstate tensions remains unexplored, although it is widely discussed as the cause of citizen-state tensions.

It is now generally accepted that dams have not delivered on 'development'. While states continue to blame neighbours for the failure of dams and 'development', international development agencies continue to encourage and fund projects on transboundary watercourses. The expansion of projects on international rivers has brought in its wake the need for mechanisms for conflict resolution between states. The need to develop a regime for managing the crisis that international river basin projects generate has been an ongoing one and parallels river basin development. The Helsinki Rules on the Uses of Waters of International Rivers 1967 attempted to deal with an important development in the post-war world, viz., the legal and institutional infrastructure for river basin development (*The Helsinki Rules on the Uses of the Waters of International Rivers* 1967). The rules envision river basin development in utilitarian terms that facilitate projects, investments and returns (D'Souza, 2002 c.). The Helsinki Rules was authored by the International Law Association, a private international association of legal professionals, on the initiative of the US Chapter of the association, largely due to the professionals in the US being interested in rules to resolve conflicts arising from river basin projects in different jurisdictions. The Helsinki Rules were objected to by participants from several developing countries (Bourne, 1996; McCaffrey, 1991). The Helsinki rules, despite their influence, are non-binding. Nevertheless, the initiatives of the International Law Association did bring the agenda of an international convention on transboundary watercourses to the International Law Commission under the UN.

The UN Convention on the Law of the Non-navigational Uses of International Watercourses 1997, initiated much after the Helsinki Rules and influenced by it, is significant for the dissidence as much as for the contents. The work of the Commission was more controversial and marked by sharp divisions. Even after years of efforts, it was not a unanimous document. Many developing countries had reservations. Burundi, China and Turkey voted against the Convention, while 27 countries abstained from voting and another 34 countries were absent.

Except France, all the countries voting against, abstaining or absenting were from the developing world, many of which were major states in international river basins (United Nations, 1997). By the time the UN Convention was finalized, the Bretton Woods organizations had altered the terms of the development discourse on water. In recent years, the World Bank has envisaged a more pro-active role for itself in mediating transboundary conflicts over water (Barrett, 1994; Kirmani and Rangeley, 1994). The effort is to save 'development' from the political tensions that it generates (Biswas, 1999).

At this meta-level, internationally, the spectres of water wars legitimate efforts by UN organs, bilateral and multilateral agencies to further develop law and institutions that will ensure river basin projects are continued and investments are safeguarded. The critique of dams and 'development' continues to be envisioned within nation-state boundaries as an issue between citizens and states. In international relations, it continues to be envisioned as a political problem between states using geopolitics as explanation. Consequently, the source of conflicts over water, the development 'project' and UN institutions, appear as the cure for water conflicts.

Re-envisioning transboundary conflicts over waters as 'development' conflicts at the meta level requires critical interrogation of the ways in which different limbs of the UN system operate to reinforce and institutionalize dominant structures of power between states and dominant discourses through which the world is envisioned. Seeing the parts without the whole insulates the legal and institutional structures of post-war capitalism and imperialism. By mystifying the legal and institutional structures of post-war capitalism and imperialism, it fetters emancipatory alternatives from emerging and leaves states open to manipulation by dominant powers in international relations. Such a 're-envisioning' is essential if we are to understand and thus resolve such conflicts in the Indian Ocean Region.

Note

1. The United Nations, Economic Commission for Asia and the Far East publication series, first under the title: 'Flood Control' later continued as 'Water Resources series' from 1949 to the present captures the dominant ideas on dams and development.

References

Alavi, Hamza. (1975). India and the Colonial Mode of Production. *Economic And Political Weekly* Special Number (August 1975):1235-1669.

Anan, Kofi. (2002). World Security Depends on Averting Water Wars - Un. *Water Engineering & Management* 149 (5):10.

Anderson, Benedict. (1983). *Imagined Communities: Reflections on the Origins and Spread of Nationalism*. London: Verso.

Arnold, David, and Ramachandra Guha, eds. (1995). *Nature, Culture and Imperialism: Essays on the Environmental History of South Asia, Studies in Social Ecology and Environmental History*. Delhi: Oxford University Press.

Arrighi, Giovanni. (1993). The Three Hegemonies of Historical Capitalism. In *Gramsci, Historical Materialism and International Relations*, edited by Stephen Gill. Cambridge: Cambridge University Press.

Bakker, Karen. (1999). The Politics of Hydropower: Developing the Mekong. *Political Geography* 18:209-232.

Barlow, Maude, and Tony Clarke. (2002). *Blue Gold: The Fight to Stop the Corporate Theft of the World's Water*. New York: The New Press.

Barrett, Scott. (1994). Conflict and Cooperation in Managing International Water Resources. Washington DC: The World Bank, Policy Research Department, Public Economics Division.

Biswas, Asit K. (1999). Management of International Waters: Opportunities and Constriants. *Water Resources Development* 15 (4):429-441.

Bourne, Charles B. (1996). The International Law Association's Contribution to International Water Resource Law. *Natural Resources Journal* 36 (Spring):155-216.

Bowett, D.W. (1964). *The Law of International Institutions*. Indian Reprint ed. New Delhi: Universal Book Traders.

Brown, Bartram S. (1992). *The United States and the Politicization of the World Bank*. London: Kegan Paul International.

Buck-Morss, Susan. (1995). Envisioning Capital: Poltical Economy on Display. *Critical Inquiry* 21 (Winter):434 - 467.

Chimni, B.S. (1993). *International Law and World Order: A Critique of Contemporary Approaches*. New Delhi: Sage Publications.

Chua, Amy L. (1995). The Privatization-Nationalization Cycle: The Link between Markets and Ethinicity in Developing Countries. *Columbia Law Review* 95 (March):223-303.

Cox, Robert W. (1993 a.). Structural Issues of Global Governance: Implications for Europe. In *Gramsci, Historical Materialism and International Relations*, edited by Stephen Gill. Cambridge: Cambridge University Press.

Cox, Robert W. (1993 b.). Gramsci, Hegemony and International Relations: An Essay in Method. In *Gramsci, Historical Materialism and International Relations*, edited by Stephen Gill. Cambridge: Cambridge University Press.

Crow, Ben. (1995). *Sharing the Ganges: The Politics and Technology of River Development*. Thousand Oaks CA: Sage Publications.

Crow, Ben, and Nirvikar Singh. (2000). Impediments and Innovations in International Rivers: The Waters of South Asia. *World Development* 28 (11):1907 - 1925.

Dirlik, Arif. (1994). The Postcolonial Aura: Third World Criticism in the Age of Global Capitalism. *The Postcolonial Aura: Third World Criticism in the Age of Global Capitalism* 20 (Winter):328-356.

D'Souza, Radha. (2001 a.). International Law, Recolonizing the Third World? Law and Conflicts over Water in the Krishna River Basin. In *Law, History and Colonialism: Empire's Reach*, edited by Diane Kirkby and Catherine Colborne. Manchester and New York: Manchester University Press.

D'Souza, Radha. (2001 b.). How the Genie of the Law Banks the Currents of the Krishna: Sharing Krishna Waters or Metamorphoses of a Colonial Problem into a Developmental Problem. Paper read at The Role of Water in History and Development, at Bergen, Norway.

D'Souza, Radha. (2002 c.). Contextualising Inter-State Disputes over Krishna Waters: Law, Science and Imperialism. Doctoral Thesis, Department of Geography, University of Auckland, Auckland, New Zealand.

D'Souza, Radha. (2002 d.). At the Confluence of Law and Geography: Inter-State Water Disputes in India. *Geoforum* 33 (2):255-269.

Economic Commission For Asia And The Far East. (1965). Proceedings of the Sixth Regional Conference on Water Resources Development in Asia and the Far East, at Bangkok, Thailand.

Economic Commission For Asia And The Far East. (1966). A Compendium of Major Interantional Rivers: Rivers in the Ecafe Region. New York: Economic and Social Council, United Nations.

Economic Commission For Asia And The Far East. (1966). The Use and Interpretation of Hydrological Data: Proceedings of Nine Seminars Conducted by Advisory Group in Afganishtan, Ceylon, China (Taiwan), Hong Kong, Iran, Malaysia, Pakistan, the Phillipines and Thailand between 22 September 1965 and 26 July 1966.

Economic Commission For Asia And The Far East. (1968). Planning Water Resources Development: Report and Background Papers of the Working Group of Experts on Water Resources Planning, August 29-September 9, 1968, at Bangkok, Thailand.

Fisher, William F, ed. (1995). *Toward Sustainable Development: Struggling over India's Narmada River*. Armonk, New York: M. E Sharpe.

Folke, Steen. (1998). Conflicts over Water and Land in South Indian Agriculture: A Political Economy Perspective. *Economic and Political Weekly* XXXIII (February 14):341-349.

Gill, Stephen. (1991). Reflections on Global Order and Sociohistorical Time. *Alternatives* 16:275-314.

Gill, Stephen, ed. (1993). *Gramsci, Historical Materialism and International Relations*. 320 vols. Cambridge: Cambridge University Press.

Gleick, Peter H. (1993). Water and Conflict: Fresh Water Resources and International Security. *International Security* 18:79-112.

Goldman, Michael Robert. (1994). "There's a Snake on Our Chests": State and Development Crisis in India's Desert. dissertation, Sociology, University of California, Berkeley.

Guha, Ramachandra. (1997). Social-Ecological Reseach in India: A 'Status' Report. *Economic & Political Weekly* Vol XXII (No 7, February 15-21):345-352.

Halliday, Fred. (1992). International Society as Homogeneity: Burke, Marx, Fukuyama. *Millennium: Journal of International Studies* 21 (3):435-461.

Harshe, Rajen. (1997). *Twentieth Century Imperialism: Shifting Contours and Changing Conceptions*. 1st ed. New Delhi: Sage Publications.

The Helsinki Rules on the Uses of the Waters of International Rivers. Adopted by the 52nd conference of the International Law Association.

Hoffbuhr, Jack W. (2002). Water Wars. *American Water Works Association Journal* 94 (2):6.

Imhof, Aviva. (1997). The Asian Development Bank's Role in Dam-Building in the Mekong. Berkeley Way: International Rivers Network.

Khan, Mohammed Shabbir. (1961). *India's Economic Development and International Economic Relations*. London: Asia Publishing House.

Kirmani, Syed, and Robert Rangeley. (1994). International Inland Waters: Concepts for a More Active World Bank Role. Washington DC: World Bank.

Kothari, Rajini. (1988). *Rethinking Development: In Search of Humane Alternatives*. Delhi: Ajanta.

Leys, Colin. (1996). *The Rise and Fall of Development Theory*. Nairobi, Bloomington & Indianapolis, Oxford: EAEP, Indiana University Press, James Currey.

McCaffrey, Stephen. (1991). International Organizations and the Holistic Approach to Water Problems. *Natural Resources Journal* 31:139-165.

McCully, Partick. (1996). *Silenced Rivers: The Ecology and Politics of Large Dams*. London & New Jersey: Zed Books.

McMichael, Philip. (1996). *Development and Social Change: A Global Perspective*. Thousand Oaks, California: Pine Forge Press.

Meier, Gerald M, and Theodore W. Schultz, eds. (1987). *Pioneers in Development Second Series*. New York: For the World Bank by Oxford University Press.

Meier, Gerald M., Dudley Seers, *et al.*, eds. (1984). *Pioneers in Development*. New York: for the World Bank by Oxford University Press.

Moore, David B., and Gerald J. Schmitz, eds. (1995). *Debating Development Discourse: Institutional and Popular Perspectives*. New York: St. Martin's Press Inc.

Shiva, Vandana. (2002). *Water Wars: Privatization, Pollution, and Profit*: South End Press.

Singh, Satyajit. (1997). *Taming the Waters: The Political Economy of Large Dams in India*. 1st ed. Delhi: Oxford University Press.

Sklar, Leonard, and Patrick McCully. (1994). Daming the Rivers: The World Bank's Lending for Large Dams. Berkeley Way, CA, USA: International Rivers Network.

Starr, Joyce R. (1991). Water Wars. *Foriegn Policy* 82 (Spring):17-36.

Taylor, Paul. (1989). The Origins and Institutional Setting of the Un Special Conferences. In *Global Issues in the United Nations Framework*, edited by Paul Taylor and A.J.R. Groom. Hampshire and London: Macmillan.

Taylor, Peter J. (1994). The State as Container: Territoriality in the Modern World-System. *Progress in Human Geography* 18 (2):151-162.

Teclaff, Ludwik A. (1991). Fiat or Custom: The Checkered Development of International Water Law. *Natural Resources Journal* 31:45-73.

Teclaff, Ludwik A. (1967). *The River Basin in History and Law*. The Hague: Martinus Nijhoff.

The World Conservation Union, and The World Bank. (1997). *Large Dams: Learning from the Past, Looking to the Future, Country*: Swiss Agency for Development and Cooperation.

Thukral, Enakshi Ganguly, ed. (1992). *Big Dams, Displaced People: Rivers of Sorrow, Rivers of Change*. New Delhi: Sage Publications.

Tvedt, Terje. (1986). Water and Politics: A History of the Jonglei Project in the Southern Sudan. Fantoft, Norway: Christian Michelsen Institute.

United Nations. (1948-49). United Nations Yearbook - 1948-49. New York: Department of Publications, United Nations.

United Nations. (1950). *United Nations Yearbook 1948-49*. New York: Department of Public Information, United Nations.

United Nations. (1951). Measures for the Ecoomic Development of under-Developed Countries. New York: United Nations Department of Economic Affairs.

United Nations. (1952). *Yearbook of the United Nations 1951*. New York: Department of Public Information, United Nations.

United Nations. (2000). *United Nations Press Release Ga/9248* [www.internationalwaterlaw.org]. United Nations 1997 [cited 16/11/2000 2000].

Washbrook, David. (1997). The Rhetoric of Democracy and Development in Late Colonial India. In *Nationalism, Democracy and Development, State and Politics in India*, edited by Sugata Bose and Ayesha Jalal. Delhi: Oxford University Press.

Willetts, Peter. (1989). The Pattern of Conferences. In *Global Issues in the United Nations' Framework*, edited by Paul Taylor and A.J.R. Groom. Hampshire and London: Macmillan.

Worster, Donald. (1983). Water and the Flow of Power. *The Ecologist* 13 (5):168-174.

CHAPTER 10

Recent Population Movements between South Asia and Australia: Trends and Implications

Graeme Hugo

Introduction

One of the most significant elements of globalization in the last two decades has been the exponential increase in the scale and complexity of international migration. As home for more than half of the world's population and three of the four most populous states, Asia has played a major role in this both as an increasingly important origin and destination of movers. In particular, south-north migration from poorer to wealthier states has assumed major significant (Massey et al., 1998). This Chapter seeks to document some of the major features of one such flow – that between India and Australia. Indian international migrations have been one of the least studied of the major world diasporas, despite its massive significance at home and abroad, and this Chapter seeks to contribute by examining the evolution in the patterns of movement between the two nations. It focuses especially on recent years. After a few comments about sources of data it briefly examines the main changes which have occurred in the flows of people between Australia and India. It then analyses recent permanent and temporary flows between the two states. It then presents a profile of the present Indian population resident in Australia and discusses some of the major policy issues relating to movement between India and Australia.

The Chapter is written largely from an Australian perspective, and it is important, at the outset, to acknowledge that international migration in

Australia has changed dramatically since the mid 1990s. Hugo (1999a) indeed has argued that Australia has entered a new paradigm of international migration, and this has seen a transformation in Australia's migration relationship with countries like India. Among the most important shifts has been the proliferation of non-permanent movements into the country, whereas, in the past, Australian governments have strongly eschewed temporary worker migration. A high proportion of people who now attempt to settle in Australia now do so as 'onshore applicants'. This includes not only those seeking to enter Australia as 'economic' or 'family' migrants but also, for the first time, Australia has become the target of substantial numbers of asylum seekers. Another trend is that more Australians than before are leaving the country on a permanent or long term basis. The government is placing increasing emphasis on skilled migration. These changes have seen a dramatic change in Australian international migration.

Some Data Considerations

Australian data on both stocks and flows of movement between Australia and India are employed in the present Chapter. These are both comprehensive and of high quality by international standards. Firstly, regarding flows, the source employed here is the Movements Data Base (MDB), maintained by the Department of Immigration, Multicultural and Indigenous Affairs (DIMIA). Each person entering or leaving Australia is required to complete an arrival or departure card containing questions on citizenship, birthplace, birthdate, gender, occupation, marital status, type of movement, origin/destination, reason (for short-term movers only) and address in Australia. This information forms the basis of the MDB, which is a high quality data base, and one of the few in the world to contain comprehensive information on both immigrants and emigrants. People leaving or coming in to Australia are classified into three types of categories according to their intended length of their stay in Australia or overseas:

- Permanent Movements
  - *Immigrants* are persons arriving with the intention of settling permanently in Australia.
  - *Emigrants* are Australian residents (including former settlers) departing with the stated intention of staying abroad permanently.

- Long-Term Movements
  - Overseas arrivals of visitors with the intended or actual length of stay in Australia of 12 months or more.

- Departures of Australian residents with intended or actual length of stay abroad of 12 months or more.
- Short-Term Movements
    - Travellers whose intended or actual stay in Australia or abroad is less than 12 months.

Clearly there are some problems associated with the use of 'intentions' as the key element in the definitions of type of movement for the MDB. It is apparent that there are no guarantees that intentions will become reality, and, as a result, there is a significant amount of category jumping which occurs (Hugo, 1994, Chapter Three). Zlotnik (1987, 933-934) has also been critical of the concept of residence used in these definitions as a 'fertile breeding ground for confusion'. Nevertheless, the MDB provides useful and comprehensive information on *flows* of people into and out of Australia which has few equals globally.

Turning to sources of information about the *stocks* of migrants, the quinquennial national censuses of population and housing are utilized. Table 10.1 shows the immigration related questions asked at Australian censuses and indicates that a comprehensive range of questions has been asked, especially in post-war censuses. Of particular interest was the introduction, in 1971, of a birthplace of parents question, which has been in each subsequent census, and the experiment with an ancestry question in 1986 and 2001. The latter has been excluded from several censuses because, although it produced a great deal of new insight into the diversity of Australia's population, it generally failed to identify third and older generations of immigrants (Khoo, 1989). Censuses have been conducted in Australia each five years since 1961 and have a low rate of under-enumeration (less than 2 percent). The census allows us to identify the first generation migrants and their Australia-born children and a number of their characteristics with a high degree of accuracy. However, the census does not provide information on former residents who have emigrated out of Australia. With respect to persons travelling out of Australia on a temporary basis, some information is obtainable if those persons left households behind who could report their absence in a question on the census schedule relating to usual residents who are absent on the night of the census. Visitors to Australia who happen to be in the country on the night of the census are counted in the *de facto* enumeration, but excluded from most data on birthplace.

The Evolution of Settler Migration from India to Australia

There is a long history of population movement between India and Australia but data on pre-war migration needs to be carefully interpreted.

Table 10.1 Ethnicity Related Topics Included in Australian Population Censuses, 1911-2001.

Topics- Persons	1911	1921	1933	1947	1954	1961	1966	1971	1976	1981	1986	1991	1996	2001
Birthplace	*	*	*	*	*	*	*	*	*	*	*	*	*	*
Birthplace of parents	*	*	*	*	*	*	*	*	*	*	*	*	*	*
Year of arrival (Period of residence in Australia)	*	*	*	*	*	*	*	*	*	*	*	*	*	*
Citizenship	*	*	*	*	*	*	*	*	*(1)	*(1)	*(2)	*	*	*
Aboriginal/TSI origin							*(3)	*(4)	*(4)	*	*	*	*	*
(Race)														
Ethnic origin											*(5)	*	*	*
Number of overseas residents or visitors								*						
Language use		*(6)	*(7)						*(8)	*(9)	*(10)	*	*	*
Religion	*	*	*	*	*	*	*	*	*	*	*	*	*	*

Notes:
(1) Prior to 1976, 'nationality' rather than 'citizenship' was asked. (2) Since 1986 the person has been asked whether or not they were an Australian citizen. (3) In all censuses prior to 1971 respondents were required to state their race and, where race was mixed, to specify the proportion of each. (4) In the 1971 and 1976 censuses a question with response categories of European, Aboriginal, Torres Strait Islander and other was included. (5) A question on each person's ancestry was asked for the first time in 1986. (6) Question asked whether the person could read and write a foreign language if unable to read and write English. (7) Question asked whether the person could read and write. (8) The 1976 census asked for 'all languages regularly used'. (9) In 1981 ability to speak English was asked. (10) Since 1986 two separate questions have been asked – Language used and ability to speak English.

This is because many of the India-born in Australia over this period were actually the children of European origin colonial officials who subsequently moved to Australia rather than being of Indian ethnic origin. Moreover, many of the people in the flows prior to the 1980s were the so-called Anglo-Indians (Waddell, 1979). Clearly, the fact that both Australian States and India were colonies of England and parts of the British Empire, and, later, the British Commonwealth meant that there were more linkages between them than Australia had with Asian countries, which were not British colonies. Nevertheless, Table 10.2 indicates that, in 1861, there were only 2,500 India-born persons in Australia. One element in the nineteenth-century movement was the recruitment of so-called 'Afghan' camel drivers from what is now Pakistan (McKnight, 1969). They were instrumental in the development of Australia's arid interior. Hence, by the time of Australia's federation in 1901, the number of India-born in Australia had trebled to 7,760. However, one of the earliest 'initiatives' of the new federal government was the introduction of a 'White Australia' policy, which effectively prevented the settlement of Indians in Australia. Accordingly, by the time of Indian independence in 1947, there were only 8,162 India-born persons in Australia.

Table 10.2. India-Born Persons in Australia, 1861-2001.

Census Year	India-Born Population	Intercensal Percent Per Annum Growth
1861*	2,500	
1901*	7,760	+2.9
1947*	8,162	+1.1
1954	11,957	+5.6
1961	14,167	+2.5
1966	15,754	+2.1
1971	29,212	+13.1
1976	37,586	+5.2
1981	42,930	+2.7
1986	47,820	+2.2
1991	61,602	+5.2
1996	77,551	+4.7
2001	95,452	+4.2

*Refers to the former British colony of India (i.e. present day India and Pakistan).
Source: Australian censuses.

The last half-century, however, has seen a substantial growth in the Indian population of Australia as Figure 10.1 indicates. The first two post-war decades saw a very slow growth of the India-born, but the bulk of the immigration was of Anglo-Indians and persons with two European-origin parents. This was a doubling of the India-born movement. However, with the breaking down of the White Australia policy in the 1960s, and its eventual total abolition in the early 1970s the tempo of Indian immigration to Australia increased, so that Australia's India-born population doubled in the 1960s and increased by 47 percent in the 1970s. This pattern of rapid growth continued in the 1980s (43.5 percent) and 1990s (55 percent).

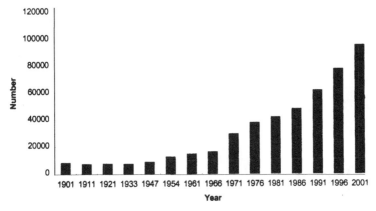

Figure 10.1. Growth of the India-Born Population in Australia, 1901-2001. (*Source*: Australian censuses 1901-2001.)

There have been, however, considerable annual fluctuations in the flow of Indian settlers to Australia as Figure 10.2 demonstrates. This shows that the numbers were quite small in the overall high immigration years of the first two post-war decades, but the release of the White Australia regulations in the late 1960s and early 1970s saw an upsurge of immigration. The overall downturn of immigration to Australia, in the mid-1970s was also experienced by Indians, but the tempo increased in the 1980s, and reached unprecedentedly high levels in the 1990s. A similar pattern is evident when we examine the proportion which Indians have made up of the total immigration inflow over the post-war period (Figure 10.3).

Increasing Indian immigration has been part of the post-war transformation of Australia from an overwhelmingly British-dominated population to a multicultural society. Table 10.3 indicates this, showing that the proportion of the national population born in dominantly non-English speaking states declined from 98.1 to 86 percent between 1947 and 2001, while those born in Asian countries increased from 0.3 to 6.5 percent. The shift which has occurred, is evident in the rates of growth of various foreign-born groups. Table 10.4 shows that India-born were among the 10 fastest-growing groups in the country in the 1990s, expanding at 4.5 percent per annum. It will be noted that virtually all of the 10 fastest-growing countries are states of the 'South', mainly from Asia. On the other hand, the slowest growing (indeed decreasing) groups were all European. In 2001-02, 5,091 Indians immigrated to Australia – the fifth largest group after New Zealand (15,663), United Kingdom (8,749), China (6,708) and South Africa (5,741).

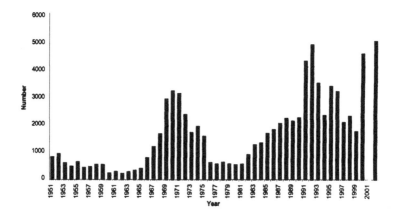

Figure 10.2. Australia: Immigrants from India, 1951-2002. (*Source*: CBCS *Demography Bulletins*; DIMIA *Australian Immigration Consolidated Statistics* and *Immigration Update*; ABS *Migration Australia*).

Note: Data from 1999-2000 comprise immigrants by country of birth. Data prior to this date comprise immigrants by country of last residence. From 1960, data are for financial years. Prior to this data are for calendar years. Data for 2001 are not available.

Immigration to Australia is a highly planned and controlled process. In the first three post-war decades, the imperatives of Australian immigration policy were both economic and demographic. On the one hand, there were massive labour shortages in the post-war boom period and labour – skilled, semi-skilled and unskilled – was needed for the massive growth in manufacturing. Also, there was a 'populate or perish' argument in the aftermath of the state being almost invaded by Japan during the war. With the end of the 'long boom' in the 1970s, the reduction in manufacturing employment and increases in unemployment, immigration policy was redefined to involve a planned numerical intake made up of a number of policy components:

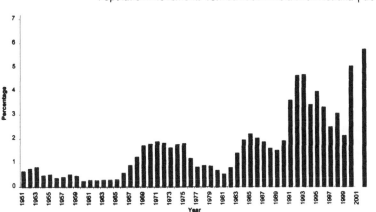

Figure 10.3. Immigrants from India as a Percentage of Total Immigrants, 1951-2002. (*Source*: CBCS Demography Bulletins; DIMIA Australian Immigration Consolidated Statistics and Immigration Update; ABS Migration Australia.)

Note: Data from 1999-2000 comprise immigrants by country of birth. Data prior to this date comprise immigrants by country of last residence. From 1960, data are for financial years. Prior to this data are for calendar years. Data for 2001 are not available.

- Refugee and Humanitarian Movement – designed to resettle refugees.
- Family Migration – enabling family members to join earlier generations of immigrants.
- Economic Migration – involving recruitment of people with skills in short supply in the economy.
- Special Categories – involving mainly New Zealanders, people with special talents, and so on.

Over the years, there has been a fluctuation in the significance of the various components of immigration (Figure 10.4). In the most recent period, there has been a deliberate policy to increase the proportion of skilled workers in the immigration intake The current Australian *Migration Program* operates within set planning levels and is made up of humanitarian and non-humanitarian components. The Skilled Migration Programme is contained within the latter and the various elements are summarized in Table 10.5. Within the program some components – that is, Business Skills, Employer Nominated Scheme (ENS), Distinguished Talents, Spouses and Dependent Children are demand driven and not subject to capping.

Table 10.3. Change in the Composition of the Australian Population by Place of Birth, 1947-2001.

	1947		2001	
	Number of Persons	Percent	Number of Persons	Percent
English speaking origin	7,438,892	98.1	15,232,338	86.0
Australia	6,835,171	90.2	13,629,685	76.9
United Kingdom and Ireland	543,829	7.2	1,086,480	6.1
New Zealand	43,619	0.6	355,765	2.0
United States and Canada	10,304	0.1	80,983	0.5
South Africa	5,969	0.1	79,425	0.4
Non-English speaking origin	140,466	1.9	2,485,110	14.0
Other Europe	109,586	1.4	1,046,967	5.9
Asia*	23,293	0.3	1,151,438	6.5
Other Africa	1,531	0.0	104,811	0.6
Other America	1,323	0.0	79,821	0.5
Other Oceania	4,733	0.1	99,361	0.6
Total	7,579,358	100.0	17,717,448	100.0

* Includes Middle East (*Source*: ABS, 1947 and 2001 censuses).

There are three main eligibility migration categories in the Migration Programme – Family, Skill and Special Eligibility. Family migration consists of a number of categories under which the potential migrant can be sponsored by a relative who is an Australian citizen or permanent resident of Australia. There has been a significant shift in recent years away from the family toward the Skilled Migration categories as is evident in Figures 10.4 and 10.5. The Skilled Migration Program consists of a number of categories of prospective migrants where there is a demand for particular occupational skills, outstanding talents or business skills. These categories are:

* Independent migrants – not sponsored by an employer or relative in Australia. They must pass a points test which includes skills, age and English language ability (22,380 visas in 2000-01).
* Skilled-Australian linked – commenced on 1 July 1997 (replacing the Concessional Family category). Applicants must pass a points test on skills, age and English ability and receive additional points for sponsorship by relatives in Australia (7,167 visas in 2000-01). Also includes Regional Linked for those sponsored by relatives in regional areas (not points tested).

Table 10.4. Australia: Fastest and Slowest Growing Groups of Foreign-Born Persons, 1991-2001.*

Country of Origin	Number of Persons 2001	Percent Growth 1991-2001
Fastest growing groups:		
Iraq	24,832	16.9
Afghanistan	11,297	15.3
Samoa	13,254	8.7
Pakistan	11,917	7.2
Korea, Republic of	38,902	6.4
China	142,781	6.1
Russian Federation	15,020	6.0
Taiwan	22,418	5.6
Thailand	23,599	5.3
South Africa	79,425	4.9
India	95,455	4.5
Slowest growing groups:		
Germany	108,220	−0.6
Cyprus	19,482	−1.3
Austria	19,313	−1.3
Malta	46,998	−1.3
Netherlands	83,325	−1.4
Spain	12,662	−1.5
Italy	218,718	−1.5
Portugal	15,441	−1.5
Greece	116,430	−1.6
Poland	58,113	−1.7
Hungary	22,752	−1.8

* Countries with 10,000 or more persons in 2001. (*Source*: ABS, 1991 and 2001 censuses.)

- Employer sponsored – employers may nominate (or 'sponsor') personnel from overseas through the Employer Nomination Scheme (ENS), Regional Sponsored Migration Scheme (RSMS) and Labour Agreements. These visas enable Australian employers to fill skilled permanent vacancies with overseas personnel if they cannot find suitably qualified workers in Australia. A total of 7,510 visas were granted in 2000-01.
- Business skills migration – encourages successful business people to settle permanently in Australia and develop new business opportunities (7,364 visas in 2000-01).
- Distinguished talent – for distinguished individuals with special or unique talents of benefit to Australia (229 visas in 2000-01).

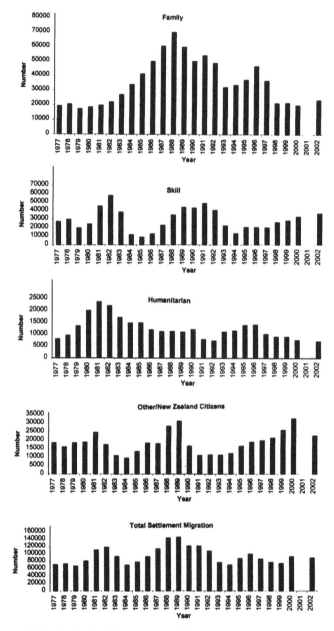

Figure 10.4. Australia: Trends in Intake of Different Types of Settlers, 1977-2002. (*Source*: DIMA, Immigration Update and Australian Immigration Consolidated Statistics, various issues.)

Figure 10.6 presents the breakdown of the numbers in each category for the year 1999-2000 to give an indication of the recent numbers in each category. Over recent times in Australia there has been greater government intervention to shape the content of the intake of immigrants so that it can better contribute to national economic development goals. This has seen greater emphasis on skills in migrant selection and in the development of Business Migration Programs involved to attract entrepreneurs with substantial sums to invest in Australia. Australia, like Canada, has micro-managed the qualifications of their migrant intake since the 1970s with the introduction of points assessment schemes.

Table 10.5. Programme Management Structure (2001-02) Migration (Non-Humanitarian) Programme.

Skill	Family	Special Eligibility
*Skilled Independent & Skilled-Australian Sponsored**	*Parents and Preferential Family* Can be capped subject to demand in	Can be capped
*Points tested	all other Family categories	
Planning level adjusted subject to demand in Business Skills and ENS	*Fiancés & Interdependents* Can be capped subject to demand for	
Business Skills, ENS & Distinguished *Talent*	spouse and dependent child places	
Demand driven	*Spouses & Dependent Children* Demand driven	
Contingency Reserve To be utilized if States and Territories, business employers and regional authorities generate additional demand, and for ICT professionals with Australian qualifications	Exempt from capping *Contingency Reserve* Legislation defeated in Senate October 2000	

*Formerly Independent and Skilled-Australian Linked (until July 1999). (*Source*: DIMIA, 2002a.)

The Skill Stream of the Australian Migration Programme is aimed at attracting people with qualifications and relevant work experience who can help to address skill shortages in Australia and enhance the size, skill level and productivity of the Australian labour force. In 2000-01, there were 44,730 people granted Skill visas, an increase of 9,400 (26.6 percent) on the 1999-2000 level. Of total Skill Stream visas, 17.1 percent (7,649) were granted to onshore applicants. The United Kingdom accounted for 15 percent of all 2000-01 Skill Stream visa grants. Other major source countries included South Africa (14 percent), India (10 percent), Indonesia (9 percent) and the PRC (8 percent).

It will be noted in Table 10.6 that Southern Asia, in which Indians are the dominant group, were over-represented in only two eligibility categories of immigration to Australia in 2001-02. These were both skilled categories. Whereas some 24.6 percent of all settlers that year were independent migrants who entered Australia via the points test, this applied to 46.6 percent from South Asia. It is clear, however, that chain migration is beginning to play an important role in South Asian migration. A quarter of all immigrants are family migrants and this is also the case for South Asia. Most interestingly, however, is the fact that, whereas 6.7 percent of all Australian immigrants were sponsored skilled settlers, some 15 percent of all South Asians were in this category.

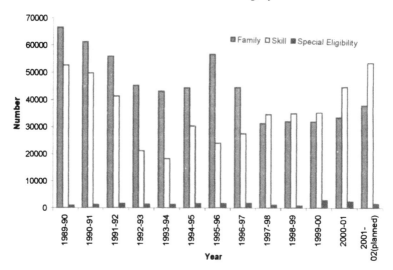

Figure 10.5. Australia: Migration Program Outcomes by Stream. (*Source*: DIMA, Population Flows: Immigration Aspects, various issues.)

The 2001 census included an ancestry question for the first time since 1986 and the results are summarized in Table 10.7. This indicated that 156,624 Australians or 0.73 percent of the total population were of

Indian ancestry. This would suggest that around 1 percent of Australians have an Indian heritage.

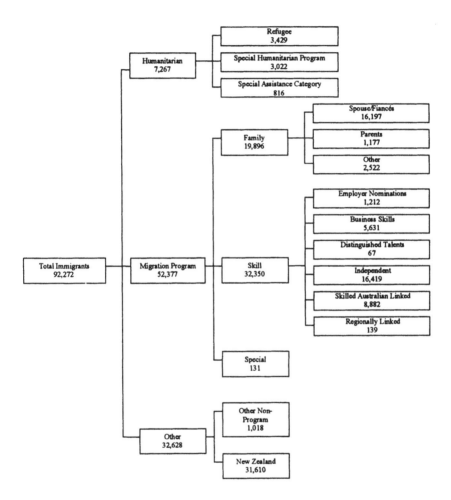

Figure 10.6 Categories of Immigration to Australia, 1999-2000, (*Source*: From data in DIMA, 2000a)

The shift in origin of settlers coming to Australia in the last three decades is demonstrated in Figures 10.7 and 10.8. The former shows that, in 1970, the dominant origin countries were European, although the beginnings of a flow from Asia, including India, is evident. On the other hand, Figure 10.8, showing the 2002 pattern, presents a quite different pattern in which Asia provides a majority of settlers.

Table 10.6. Settler Arrivals Born in Southern Asia Compared with Total Intake According to Eligibility Category, 2001-02.

Eligibility Category	Southern Asia Number	Total Number	Percent of Total from Southern Asia
Family	2,165	23,344	9.3
Skill			
Sponsored	1,374	5,960	23.1
ENS	42	1,817	2.3
Business	119	6,409	1.9
Independent	4,284	21,850	19.6
Special Eligibility	1	176	0.6
Humanitarian Programme	611	6,732	9.1
Non-Programme Migration			
. NZ Citizen	565	21,458	2.6
Other	29	1,154	2.5
Total	9,190	88,900	10.3

Source: DIMIA, 2002b, 7.

Fgure 10.7. Distribution of Birthplace of Settlers to Australia, 1970. (Source: DIMA, Immigration Update, various issues.)

Non-permanent Migration

An increasing proportion of Australia's skilled migrant workers involves onshore approval with persons entering Australia under some other visa category and then applying to settle in the country. This process has been made easier in recent times for some groups. For example, overseas students who graduate from Australian universities can readily gain entry with their qualifications not having to be assessed. The Assessment of Qualifications of immigrants is an issue of concern. The degree to which there is a formal necessity to have qualifications assessed varies with the type of skill of the immigrant. In some areas (for example, information technology) there is little professional regulation, while, in others (for example, doctors), there is a high level of professional assessment and regulation.

Table 10.7. Ancestry of the Australian Population, 2001.

	Number	Percent
Oceanian		
Australian	6,739,595	31.46
Other Australian Peoples	106,454	0.50
Maori	72,959	0.34
New Zealander	123,328	0.58
Other Oceanian	91,727	0.43
North West-European		
English	6,358,882	29.69
Scottish	540,043	2.52
Irish	1,919,723	8.96
Dutch	268,754	1.25
German	742,210	3.47
Other NW European	346,441	1.62
Southern and Eastern European		
Italian	800,257	3.74
Maltese	136,755	0.64
Croatian	105,745	0.49
Greek	375,699	1.75
Macedonian	81,893	0.38
Serbian	97,326	0.45
Polish	150,903	0.70
Russian	60,213	0.28
Other SE European	353,646	1.65
North African and Middle Eastern		
Lebanese	162,245	0.76
Turkish	54,597	0.25
Other North Africa and Middle Eastern	147,030	0.69
South-East Asian		
Vietnamese	156,572	0.73
Filipino	129,831	0.61
Indonesian	28,265	0.13
Other SE Asian	91,316	0.43
North-East Asian		
Chinese	556,553	2.60
Other NE Asian	80,676	0.38
Southern and Central Asian		
Indian	156,624	0.73
Other Southern and Central Asian	139,223	0:65
Peoples of the Americas	140,121	0.65
Sub-Saharan Africa	103,750	0.48
Total	21,419,356	100.00

Note: Table excludes inadequately described, not stated and not applicable. (*Source*: ABS, 2001 census.)

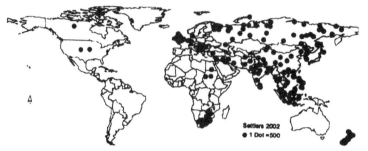

Figure 10.8. Distribution of Birthplace of Settlers to Australia, 2002. (*Source*: DIMIA, 2002b.)

In post-war Australia, there has been bi-partisan agreement that permanent settlement of a significant number of overseas migrants is desirable, and, accordingly, each post-war government has had an active Immigration Programme while the non-permanent labour migration encouraged by some other OECD states during this period was strongly opposed. There has been no challenging the notion that the permanent settlement paradigm dominated Australian thinking and policy regarding international migration. There has, however, been a change (Hugo, 1999a). In response to major changes in the Australian economy, internationalization of labour markets and globalization forces more generally, there has been a major shift in policy, which has allowed entry of large numbers of people who have the right to work in Australia on a non-permanent basis. Figure 10.9 depicts recent trends in the major non-settlement categories under which people may enter Australia and be allowed to work. It is argued elsewhere that this represents a parametric change in Australian immigration policy (Hugo, 1999a). Nevertheless, this type of visa entry has not been extended to unskilled and low skilled areas, and has been strictly open only to people with skills in demand and entrepreneurs.

A major watershed in Australian post-war immigration policy occurred in 1995 when the Labor government received a report headed by Mr.Neville Roach which recommended that the government:

'fundamentally deregulated the rules governing temporary work visas for skilled business persons and specialist workers (mainly professional and managerial level workers, though trade level workers are eligible too)' (Birrell, 1998, 1).

As a result, in 1996, the new Coalition government put in effect most of the recommendations of the report:

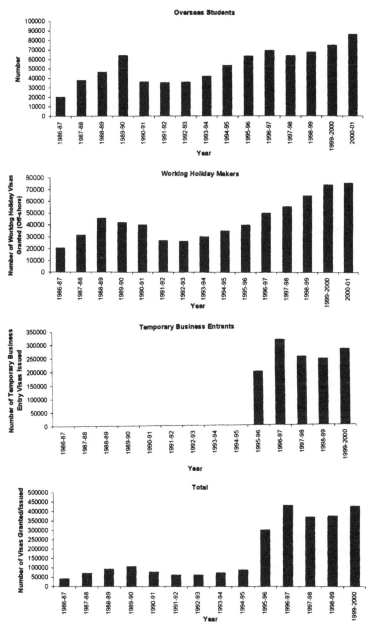

Figure 10.9. Non-Permanent Migration to Australia of Persons with the Right to Work by Category, 1986-2001. (*Source*: DIMA, *Population Flows: Immigration Aspects*, various issues; Ruddock, 2002.)

- Removal of previous restrictions governing the sponsoring of business persons and specialists.
- Abolition of previous requirements that sponsors establish that there are no resident Australian workers available to do the work.
- Abolition of the requirement that there would be some training benefit to Australian workers.
- Liberalization of rules governing the sponsorship process.
- Establishment of a pre-sponsoring arrangement whereby employers can register themselves as sponsors if they meet minimal requirements. They then are allowed to sponsor any number of the new 457 Temporary Entry category.
- A two stage process was introduced for the entry of 457 workers:
  – the sponsor nominates the position;
- – if there is no objection from DIMA, the applicant can apply for a 457 visa.

The Temporary Business Visa category was introduced in 1996 and as DIMA (2000b, 48) points out:

> The employer sponsored temporary business visas allow employers to fill skill shortages from overseas and assess new ideas, skills and technology. The visa holders tend to be highly skilled and have relatively high income levels and therefore able to contribute to economic growth through improved productivity and increased demand for goods and services. The entry of managers and skilled specialists under these categories can also enhance Australia's ability to compete in international markets.

One of the distinguishing features of the exponential increase in non-permanent migration to Australia is that the pattern of the origin countries of groups who are granted temporary residence and the right to work is quite different to that of permanent settlers. Figure 10.10 shows the origins of the 108,171 persons coming to Australia under the Working Holiday Maker Programme. This program is restricted to a limited range of countries, predominantly in Europe, and this is reflected in the origins of this group. A second temporary entry group, which has greatly increased in significance, is that of students. Australia has a greater ratio of overseas to local students in its tertiary education institutions than any other OECD state.

In 2001-02, 223,600 students arrived in Australia to study, 88,853 for the first time. Of the former, 6,452 (2.9 percent) and 2,696 (3 percent) of the latter were from India. Australia has embraced a policy of selling education, especially to the Asian region, so that it is now among the country's five largest export earners. Figure 10.11 shows that the bulk of these students are drawn from Asian states. Of the overseas students resident in Australia in mid 2002, 78.9 percent came from Asia. However, the bulk of these are from Southeast and North Asia, with India contributing only 2.7 percent of the total overseas student numbers.

The origin of temporary resident arrivals, which includes long-stay business, working holiday makers and occupational trainees, but excludes students and visitors, is depicted in Figure 10.12. While there is a significant group from Asia, more developed regions account for the majority. Table 10.8 shows that 53.1 percent came from Europe and the USSR, 8.4 percent from Northern America and 10.8 percent from Japan. The Indian representation among temporary residents (2.9 percent) is significant, but lower than their representation among the total resident overseas-born population.

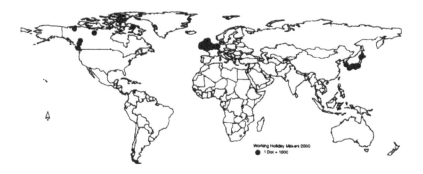

Figure 10.10. Australia: Working Holiday Makers Arrivals, 2000. (*Source*: Drawn from data in DIMA, 2001)

Figure 10.11. Australia: Student Arrivals, 2000. (*Source*: Drawn from data in DIMA, 2001.)

Figure 10.12. Australia: Temporary Resident Arrivals, 2000. (*Source*: Drawn from data in DIMA, 2001.)

Table 10.8. Australia: Origins of the Australian Foreign-Born Total and Temporary Resident Populations, 2001 and 2002.

	Total Overseas-Born, 2001		Temporary Resident Overseas-Born, 2002	
	Total	Percent*	Total	Percent*
Oceania	455,126	11.1	1,892	1.7
Europe and Former USSR	2,136,158	52.3	60,563	53.1
Middle East and North Africa	213,942	5.2	1,854	1.6
Southeast Asia	497,076	12.2	6,289	5.5
Northeast Asia	298,831	7.3	22,181	19.5
Southern Asia	184,130	4.5	4,577	4.0
India	95,452	2.3	3,356	2.9
Northern America	81,403	2.0	9,616	8.4
Southern America	75,691	1.9	984	0.9
Africa	141,696	3.5	6,044	5.3
Not stated	1,051,805**	-	16,053	-
Total	5,135,858	100.0	130,053	100.0

*Excluding 'Not stated'.
**Includes 'Inadequately described', 'At sea' and 'Not elsewhere classified'.
Source: ABS 2001 census and DIMIA.

It is apparent that India has participated in the substantial expansion of temporary migration to Australia, although not to the same extent as some other groups. However, Table 10.9 shows that there has been a major increase in the number of India-born persons visiting Australia on a long-term basis. The numbers doubled between 1992 and 1996, between

1996 and 1998, and, again, between 1998 and 2000. This reflects the increased involvement of Indians in the influx of skilled temporary workers to Australia. They have been heavily involved in the influx of information technology workers for example. It will be noted in the Table that not only have the numbers entering under the Temporary Entry category increased, but the net migration gain has also increased. This has been a feature of Australian population change in recent years. Net migration has accounted for around half of the national population increase, but half of that net gain has been in temporary movement. It remains to be seen how much of the temporary migration will see category jumping into permanent settlement. Some strong indications of this increasingly occurring is the substantial increase in the number of Australian settlers applying for settlement onshore (28.1 percent in 2001-02).

Table 10.9. Australia: Long-Term Movement To and From India and Australia, 1994-2000.

	Long-Term Resident			Long-Term Visitor			Total Long Term Net Migration
	In	Out	Net	In	Out	Net	
1994	607	510	+97	915	563	+352	+449
1995	704	550	+154	1,453	701	+752	+906
1996	696	626	+70	2,441	1,006	+1,435	+1,505
1997	712	694	+18	2,903	1,596	+1,307	+1,325
1998	711	699	+12	4,461	2,300	+2,161	+2,173
1999	631	687	−56	5,321	2,330	+2,991	+2,935
2000	718	643	+75	6,048	3,459	+2,589	+2,664

Source: ABS, *Migration Australia*, various issues.

Table 10.10. Australia: Short Term Movement To and From India and Australia, 1995-2000.

	Short Term Resident		Short Term Visitor	
	In to Australia	Out of Australia	In to Australia	Out of Australia
1995	24,194	23,454	25,858	25,236
1996	24,816	28,386	33,159	31,203
1997	31,166	27,758	37,470	36,479
1998	30,107	31,765	44,618	42,523
1999	35,660	31,624	49,765	47,345
2000	34,951	35,374	58,331	57,716

Source: DIMIA Movements Data Base.

It is also interesting to note shifts in short-term movement in recent years. Table 10.10 shows that there has been an increase in both Australian movement to India and that in the opposite direction. In particular, Indian short-term movement (involving primarily tourists and short-term business visitors) to Australia has increased substantially – doubling between 1995 and 2000. In 1995, there were similar numbers of Australians visiting India or Indians visiting Australia. However, by 2000, there were almost twice as many Indians visiting Australia as Australians going in the other direction.

Movement From Australia to India

There is a tendency for Australia to be categorized as a purely immigration country, but, in fact, it also is a country of significant emigration. In recent years, departures on a long-term or permanent basis have been very substantial in relation to the intake of immigrants. Former settlers have formed a major part of the outflow, as Figure 10.13 indicates. In 1999-2000, permanent and long-term departures (197,846) reached unprecedented levels and the proportion which was Australian-born among the permanent departures reached its highest ever levels (49.3 percent).

Table 10.11 provides occupational details of those leaving Australia permanently. It will be noticed that, although Australia receives a net gain of all occupational categories, the occupational profile of emigrants is somewhat higher than that of the permanent arrivals. The main difference is in the highest status manager/administrator category, which accounts for 18.2 percent of the emigrants, but only 12 percent of the immigrants. It will also be noted that 61.7 percent of the emigrants were in employment before moving, compared with 49.8 percent of settler arrivals.

Settler loss has been an important feature of the post-war Australian migration scene with around a fifth of all post-war settlers subsequently emigrating from Australia, most of them returning to their home country. There has been concern about this settler loss among policy-makers (Hugo, 1994), but it has a number of components including a group of migrants who never intended to settle permanently in Australia as well as people who are influenced by family changes, those unable to adjust to life in Australia, and so on. The pattern of settler loss, while it varies between birthplace groups (for example, it is high among New Zealanders but low among Vietnamese), has tended to remain a relatively consistent feature of the post-war migration scene in Australia and the fluctuations in its numbers are very much related to earlier levels of immigration. With an increase in the skill profile in immigration, we can expect an increase in settler loss, since skilled migrants have a greater chance of remigrating than family migrants.

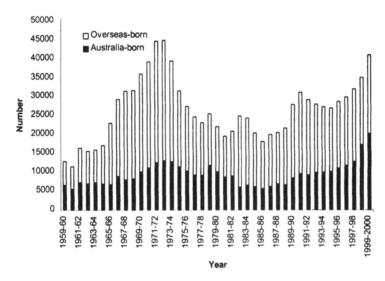

Figure 10.13 Permanent Departures of Australian-Born and Overseas-Born Persons from Australia, 1959-60 to 1999-2000. (*Source:* DIMA, Australian Immigration Consolidated Statistics and Immigration Update, various issues.)

A rough indication of contemporary patterns of settler loss can be derived from comparing the birthplace of permanent arrivals and departures to and from Australia. These data are presented in Table 10.12. This indicates that there are relatively high rates of return among settlers from more developed countries in New Zealand, Europe and North America. There are also high rates of return to parts of Northeast Asia, especially Japan (ABS, 2001). However, there is every indication of a low rate of settler return among South Asian arrivals.

It will be noted in Table 10.12, however, that more than a half of the permanent departures from Australia in 2001-02 were of Australia-born persons. While a small number of these were the Australia-born children of former settlers, there can be no doubt that the numbers of Australia-born persons leaving the country on a permanent basis is increasing as Table 10.13 indicates. Focusing on the Australia-born movement to India, Table 10.14 indicates that, while there has been an increase in the flow it is still relatively small in number, especially when compared with the UK and the USA, which together are the destination of half of all Australia-born long term and permanent out-movement.

Table 10.11 Australia: Percent Arrivals and Departures 1999-2000 by Occupation

Occupations	Settler Arrivals		Permanent Departures		Difference
	Number	Percent	Number	Percent	
Managers & administrators	5,519	12.0	4,605	18.2	+914
Professionals	17,065	37.1	8,965	35.4	+8,100
Associate professionals	4,788	10.4	2,899	11.4	+1,889
Tradespersons	6,075	13.2	1,844	7.3	+4,231
Advanced clerical & service	1,395	3.0	990	3.9	+404
Intermediate clerical & service	5,487	11.9	3,483	13.7	+2,004
Intermediate production & transport	1,525	3.3	555	2.2	+970
Elementary clerical, sales, service	2,638	5.7	1,458	5.8	+1,180
Labourer & related workers	1,453	3.2	532	2.1	+921
Total workforce	45,945	100.0	25,351	100.0	
Total in employment		49.8		61.7	
Not in employment	4,134	4.5	569	1.4	+3,565
Not in labour force	41,228	44.7	15,079	36.7	+26,199
Not stated	965	1.0	79	0.2	+886
Total	92,272	100.0	41,078	100.0	+51,194

Source: DIMA, 2000a

Australia-born emigration has begun to attract policy attention since the profile of departures of residents tends to be younger and more educated than the population of the country as a whole and the spectre of a 'brain drain' has arisen. There can be no doubt that the global international migration system with respect to highly skilled labour has been transformed since the 1960s, when the first 'brain drain' research was undertaken. Findlay (1990, 15), however, has summarized the situation as follows:

professional transients are replacing settler migrants in the international migration systems of many countries. Where settler emigration is still permitted, it is skilled migrants who find it easiest and quickest to receive work and residence permits . . . Most nations, however, do not continue to favour large-scale settler migration and instead seek to meet specific skill shortages by permitting (if not promoting) transient skill movements. The transient skill flows already dominate the migration systems of some countries such as the United Kingdom . . . They involve the international circulation of high level manpower between countries, with the migrants neither seeking nor being encouraged to remain in any particular place for a long time period. Foreign assignments are commonly for one to three years . . . Given the circulatory nature of these high level manpower movements, it has been suggested that these migration moves be seen as skill exchanges rather than brain drain.

Table 10.12. Australia: Birthplace of Permanent Arrivals and Departures, 2001-02.

Region	Departures		Arrivals		Ratio
	Number	Percent	Number	Percent	In:Out
Australia	24,146	50.1	–	–	
Other Oceania	6,398	13.2	19,152	21.5	2.99
Europe and former USSR	6,529	13.5	17,411	19.6	2.67
Middle East	753	1.6	6,000	6.7	7.97
Southeast Asia	2,641	5.5	14,464	16.3	5.48
Northeast Asia	5,548	11.5	10,716	12.1	1.93
Southern Asia	424	0.9	9,190	10.3	21.67
Northern America	1,051	2.2	1,730	1.9	1.65
Southern America	263	0.5	900	1.0	3.42
Africa	481	1.0	9,311	10.5	19.36
Not stated	7	–	26	–	3.71
Total	48,241	100.0	88,900	100.0	

Source: DIMIA, 2002b.

Whereas, in the 1960s, the dominant form of professional international migration tended to involve permanent migration from less developed to more developed states, the current situation tends to be characterized by the transilience of such groups; that is, hypermobility involving remigration and return (Richmond, 1991, 4). Seen in this context, it is somewhat unrealistic to expect that Australia will be isolated from this process and have its international migration of skilled groups dominated by traditional settler emigration movements. The greatly increased significance of transilience in skilled labour movements has been encouraged by a number of developments over the last decade or so:

- Many highly skilled regional and national labour markets have been usurped by labour markets which overlap international boundaries.
- The internationalisation of capital.
- The exponential development of exchanges of all types through the development of communications.
- The reduction in real time and money costs of travel.
- The development of multinational corporations.

Table 10.13. Australia: Permanent Departures, 1991-2002.

Year	Former Settlers	Australia-Born	Total
1991-92	19,944	9,178	29,122
1992-93	18,102	9,803	27,905
1993-94	17,353	9,927	27,280
1994-95	16,856	10,092	26,948
1995-96	17,665	11,005	28,670
1996-97	18,159	11,698	29,857
1997-98	19,214	12,771	31,985
1998-99	17,931	17,250	35,181
1999-2000	20,844	20,234	41,078
2000-01	na	na	na
2001-02	24,095	24,146	48,241

Source: DIMIA, Immigration Update, various issues.

Australia's Indian Community

The impact of any migration stream on a destination society goes far beyond the demographic addition of new residents. Migrants always have a different socio-economic and demographic profile different from that of populations at the origin and destination, so they will change those populations through their presence or absence. Moreover, migrants often provide new channels for the flows of information, goods and capital. It is important, therefore, to examine the characteristics of Indians living in Australia.

One of the universal features of migration streams is that they tend to be selective of particular age groups, especially young adults. That this is also the case for Indian migration to Australia is evident in Figure 10.14 which overlays the age-sex pyramid of the Indian population recorded at the 2001 census who had been resident in Australia less than 10 years with that of the total Australian population. It indicates a dominance of young adults aged between 20 and 44 (males) and 25-40 (females). There is a significant under-representation of dependent children and older persons. This, of course, partly reflects the fact that age is one of the criteria upon which migrants are selected into or out of the Immigration Program. However, if the total Indian population is compared with the total Australian population (Figure 10.15) the Indian population is somewhat older than the total. This reflects the fact that, as immigrant populations age, they contribute to the overall ageing of the population so that immigration only makes the total population younger, if it is continually increasing in size (Young, 1988).

Table 10.14. Permanent and Long Term Outmovement of the Australia-Born Who Went to India, 1993-2000.

Year	Total	Percent
1993-94	536	0.7
1994-95	584	0.7
1995-96	651	0.8
1996-97	719	0.8
1997-98	738	0.8
1998-99	722	0.7
1999-2002	680	0.6

Source: DIMIA Movements Data Base.

One of the most significant demographic events occurring in Australia in the 1980s was that, for the first time in the country's post-European history, females outnumbered males. Despite the fact that women outlive men in Australia to a greater extent than most countries (Hugo, 1986), the dominance of males has until recently more than counterbalanced the effect of the longevity difference. However, the last two decades have seen a feminization of settlement in Australia. For example, in 2002-02, there were 93.7 males for every 100 females arriving in Australia. While females are especially important in the stream from Asia, there are more males than females in the migration from India. Hence, at the 2001 census, the sex ratio among India-born was 112.0.

Migration is often selective of the more adventurous, entrepreneurial, better trained and risk-taking populations. Moreover, this is exacerbated in highly planned immigration programs, like that of Australia, which has become increasingly focused on selecting settlers on the basis of skill (Richardson, Robertson and Ilsley, 2001). This is reflected in the characteristics of the national India-born population (Hugo, 1999b). Table 10.15 shows that the Indian population of Australia has a higher proportion in the managerial, administrative and professional

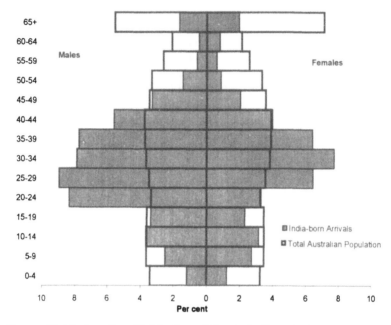

Figure 10.14. Age-Sex Distribution of the India-Born in Australia Arriving From 1991 to 2001 Compared With the Total Australian Population in 2001. (*Source*: ABS 2001 census.)

occupation categories than the total workforce, and almost three times a greater proportion with a University degree. One problem they have suffered, however, has been a high level of unemployment than the total population. At the 1996 census, 9.6 percent of Indians in the workforce were unemployed, compared with 8.6 percent of the Australia-born. Recent longitudinal research of the labour market experience of two cohorts of immigrants arriving in Australia (1993-95 and 1999-2000 arrivals) in which Indians were included, has shown that the later cohort, in which there was a greater emphasis on skill in selection, had greater success in entering the labour market (Richardson, Robertson and Ilsley, 2001). Nevertheless, in July 2002 (DIMIA, 2002b, 34), the India-born population had an unemployment rate of 7.5 percent, which was above those of the total overseas-born (6 percent) and the Australia-born (5.5 percent). There is considerable anecdotal evidence that skilled Indians experience difficulty in entering occupations commensurate with their qualifications and the jobs they held in India. Problems of recognition of qualifications are an issue in the population. One aspect of Indian migration to Australia which cannot be disputed is that there is a definite pattern of brain drain in the immigration from India to Australia. Birrell et al. (2001, 17) show that the net gain of persons with professional occupations over the 1997-98 and 1999-2000 period was 32,628, and 7,800 of that net gain was of Indians. They were the fifth-largest contributors to that gain after New Zealand, South Africa, China and the UK-Ireland.

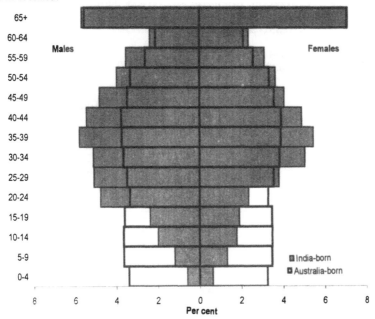

Figure 10.15. Age-Sex Distribution of the India-Born Population and the Total Australian Population in 2001. (*Source*: ABS 2001 census.)

The positive selectivity of Indian immigration to Australia is reflected in the fact that the community has on average a higher socio-economic level than the total Australian population. This is evident in Table 10.16, which shows that, despite their recency of arrival, the proportion of Indians in the highest income category is almost twice that of the Australia-born. Also, despite their recent arrival, the rate of owner-occupancy of housing among them is almost as high as for the Australia-born.

Patterns of Settlement in Australia

Most immigrant groups in Australia have a quite different spatial distribution than the Australia-born, reflecting the pattern of job opportunities prevailing at their time of arrival in Australia and the existence of communities of prior arrivals. Where immigrants settle can have a major impact on their level of satisfaction. Indians have had a quite distinctive pattern of settlement in Australia. This is reflected at a macro-level in Table 10.17. This shows that 90.2 percent of Australian Indians live in the 8 capital cities compared with 58.2 percent of the Australia-born and 80.5 percent of the total overseas-born. This metropolitan concentration has tended to increase over recent years. In 1991, 87.7 percent lived in capital cities.

Table 10.15. Australia: Australia- and India-Born: Selected Occupational and Educational Characteristics, 1996.

	Birthplace		Total Overseas-Born %	Total Population %
	Australia %	India %		
Occupation:				
Managers/Admin./ Prof.	27.2	36.0	27.5	27.2
Technician/Assoc. Prof.	11.5	11.4	11.9	11.6
High Skill-Trade Clerical	17.9	13.1	17.5	17.8
Intermediate Skill	25.4	25.8	25.1	25.3
Low Skill-Clerk Labourer	18.0	13.7	18.0	18.1
Percent Unemployed	8.6	9.6	10.7	9.2
Highest Qualification:				
Degree/Diploma	16.2	42.7	19.2	16.5
Skilled/Basic Vocational	14.1	9.3	13.1	13.3

Source: 1996 Australian Census of Population and Housing.

One of the most distinctive features of the historical settlement of Indians in Australia has been what Hugo (1996) has described as an 'Indian Ocean Connection'. He has shown how data from the 1991 census showed that the ring of states around the Indian Ocean has a disproportionately large share of its Australian populations living in the State of Western Australia. In 1981, Western Australia had almost a quarter (24.3 percent) of the Australian Indian population – more than three times that State's share of the national total population. This is obviously partly a factor of greater proximity to the Indian Ocean region than any other Australian State, the fact that Perth was the first port of call of early settlers from India, and, that the establishment of substantial Indian communities in Perth have served as anchors for later settlement (Waddell, 1979).

Table 10.16. Australia:Australia- and India-Born: Selected Socio-economic Indicators, 1996.

	Birthplace		Total Overseas-Born %	Total Population %
	Australia %	India %		
Nature of Occupancy:				
Owner/Purchaser	72.0	65.5	69.5	71.3
Tenant	27.1	33.6	29.6	27.8
Other	0.9	0.9	0.8	0.9
Income:				
Less than $300 per week	49.8	43.3	52.8	50.8
$800 per week or more	10.9	16.5	10.8	10.8

Source: 1996 Australian Census of Population and Housing.

Table 10.17. Australia: Australia- and India-Born: Spatial Distribution, 2001.

	Birthplace		Total Overseas-Born %	Total Population %
	Australia %	India %		
Urban/Rural Distribution:				
Capital City	58.2	90.2	80.5	63.3
Rest of State	41.8	9.8	19.5	36.7
Interstate Distribution:				
New South Wales	32.7	39.7	36.3	33.6
Victoria	24.0	32.1	26.6	24.5
Queensland	20.4	7.7	15.2	19.3
South Australia	8.1	3.9	7.3	7.7
Western Australia	9.1	13.7	12.2	9.8
Tasmania	2.8	0.5	0.2	2.4
Northern Territory	1.2	0.5	0.5	1.1
ACT	1.7	1.9	1.6	1.6

Source: 2001 Australian Census of Population and Housing

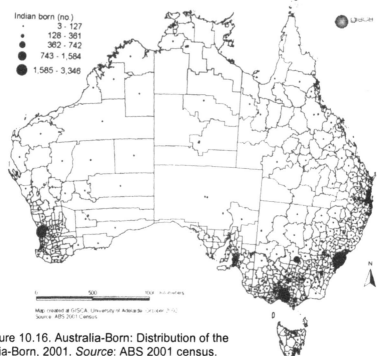

Figure 10.16. Australia-Born: Distribution of the India-Born, 2001. *Source*: ABS 2001 census.

However, the last two decades have seen an eastward shift in Indian settlement in Australia. Each recent census has seen an eastward shift in the focus of settlement. Hence, by 2001, the proportion of Australian Indians living in Western Australia had fallen to 13.7 percent, still well above its share of the total Australian population (9.1 percent) but half its share a quarter-century earlier. Like most Asian groups, the largest single concentration of Indians in Australia is in Sydney, but the Melbourne community is almost as large. More than three quarters of Indians arriving in Australia now go initially to Sydney and Melbourne. They are heavily underrepresented elsewhere, especially in Queensland.

Figure 10.16 depicts the distribution of the India-born in Australia at the 2001 census and this shows the concentration in capital cities, especially Sydney, Melbourne and Perth. One much discussed aspect of immigrant settlement in Australia is the degree to which particular groups tend to concentrate spatially in communities within the country's major cities. Indians have shown only a minor tendency to concentrate, perhaps reflecting their high socio-economic status and ability to communicate effectively in English. Table 10.18 shows that, in Sydney, around a quarter of the India-born would have to change their statistical local area of residence to duplicate the distribution of the Australia-born, indicating a moderate degree of spatial concentration. It is interesting that the tendency to concentrate has increased slightly over recent years.

Table 10.18. Sydney Statistical Division: Indices of Dissimilarity, India-Born, 1981-96.

Year	Index of Dissimilarity
1981	22.5
1986	18.8
1991	23.2
1996	27.6

Source: Calculated from ABS 1981, 1986, 1991 and 1996 Censuses.

International Migration Issues in Australia

As one of the world's few 'traditional migration' countries, international migration has been at the forefront of public debate and policy interest over a long period of time in Australia. However, these issues have never been so dominant as they are at present. There is only space here to briefly raise some of the major issues, but several impinge upon Indian migration and settlement.

Migration and Demographic Growth

Although Australia has higher levels of fertility[2] than many OECD states, they are substantially below replacement level and the outlook is for natural increase to slow down and cease in around 25 years, if there is zero net immigration. Moreover, their populations are ageing, so that the ratio between the working age groups and the dependent older population is becoming less favourable. While the situation is by no means as dramatic as in European states, the issue of whether immigration can be used as 'replacement migration' (United Nations, 2000) to help achieve a more balanced age structure when population stability is achieved is gaining increasing attention. Of course, there are proponents of greatly increasing immigration to expand national populations because it is seen as enhancing economic and political status. However, there are also those who point out the substantial environmental limits and advocate limiting immigration to current levels or reducing it.

Migration and Social Cohesion

Australia entered the post-World War II era as states which were overwhelmingly Anglo-Celtic, despite its indigenous populations. Immigration has changed this dramatically so that it has become a multicultural state. In Australia, there are 53 countries of birth with more than 10,000 people living in Australia (Hugo, 1999b). This transformation has been achieved largely without the conflict which has characterized some instances where incoming immigrants differ substantially in culture, language, religion and so on, than the dominant population. Of course, there are still elements who oppose multiculturalism in Australia (for example, see Jones, 1998) but it is official policy and embedded in the legal system.[2] Sydney now has 33.5 percent of its population born overseas. Currently, 5.5 percent of the national population was born in Asia, 11.3 percent of Sydney's population were born in Asia. These represent a major change and continue to be an area of debate. However, it must be said that immigration in Australia does not provoke the negative reactions which have been seen over the last decade over much of Europe and Japan. This is because a majority of the people have witnessed large-scale immigrations and the positive effects from those movements.

Migration and Economic Growth

Much of the discussion around immigration in contemporary Australia centres around the issue of its effects on the economy. As with other OECD states, immigration is currently seen as an important mechanism to recruit skilled labour in areas of the economy which are currently constrained by the lack of locally-trained human resources. Competition for immigrants with skills in areas such as engineering, information technology, management, research, etc. between the OECD states is leading to the development of new elements in immigration programs. There are debates about these policies, which include:

- The 'brain drain' effects of recruiting skilled workers from less developed countries on their home countries' development.
- The extent to which this is being used as a substitute for building up national training and education capacity to produce the skills needed in the 'New Economy'.
- Problems with the recognition of skills and qualifications obtained overseas.
- The extent to which immigrants coming in under 'business' migration programs are in fact able to transfer their business and entrepreneurial skills to new contexts which represent quite different business environments to those in Asia.

There are strong pro-immigration lobby groups within both countries that argue that population growth and economic growth are positively correlated. These especially include groups such as the housing lobby for who the economic benefits of immigration are immediately evident. Immigrants continue to experience greater difficulty than non-immigrants to enter the labour market and rely on unemployment benefits, although research in Australia would suggest that the recent changes in immigration policy in sharpening its economic focus have reduced these effects (Richardson et al., 2001). There have also been arguments that increased immigration is an evitable corollary of the increasing of other flows (trade, finance, tourism and so on) which are involved in globalization and are part of national efforts to embed the economy more into that of the fast developing Asia-Pacific region.

Refugee and Humanitarian Migration

Australia has a strong and sustained tradition of taking insignificant numbers of refugees for permanent settlement. However, until recently, this has involved the selection of refugee settlers from persons undergoing determination of their refugee status outside of the country. This changed with increased numbers of asylum seekers who either arrive in airports or on boats on the northern coast of Australia. It became an issue of substantial significance in Australia with the government, for the first time in September 2001, not allowing boatloads of asylum seekers to land in Australia and diverting them to other Pacific countries to have their refugee status determined. Australia has a policy of mandatory detention of asylum seekers until their claim for refugee status is heard and this has attracted considerable controversy (Hugo, forthcoming). It is a very complex issue with some characterizing these people as 'illegal migrants', 'queue jumpers', and linked with organized crime through people-smuggling, and others seeing them as genuine refugees. Of course, both elements are in the flow. In 2000, 45.8 percent of the boat people whose applications for determination had been held had been given status to allow them to settle in Australia.

Undocumented Migration

The island status of both Australia and New Zealand has tended to protect them from having substantial influxes of undocumented migrants, at least of the kind who clandestinely enter the country. Undocumented or illegal migration to Australia and New Zealand is of three types:

- Overstaying whereby non-citizens enter legally but overstay the term of their visa (overstayers).

- Where non-citizens entering legally otherwise ignore the terms of their visa, e.g. persons on a tourist visa working.
- Clandestine entry of non-citizens who do not pass through an immigration control point (illegal entrants).

Much is known in Australia about overstayers, since there is a high quality Movement Data Base, and all persons arriving in and departing from the country are required to complete a card which facilitates matching and detection of overstayers. Through the 1990s around 50,000 overstayers have been identified using this matching. In December 2000, there were 58,674 overstayers of whom 28 percent had been in Australia for more than 9 years and a similar proportion had been in the country for less than a year (28.6 percent). Some 79.5 percent of overstayers were persons who had overstayed tourist visas, 5.4 percent temporary residents and 7.5 percent students. The overstay rate was 0.2 percent, comprising 7,196 overstayers from 3,848,993 visitors in 1999-2000 (DIMA, 2000b, 56). It is estimated that there are 1,478 overstayers from India in Australia (DIMIA, 2002a).

Turning to the people who enter Australia illegally, it is clear that Australia has, in recent times, become a more important target for such movements. There are, of course, no data on persons who have been successful in such attempts but there are on the numbers who have been detected. These can be divided into those detected arriving by air and those coming by boat. There has been a substantial increase in the numbers detected in recent years. In 1998-99, 2,106 people were refused entry at Australia's airports (an increase of 35.9 percent over the previous year's 1,550). However, it subsequently fell to 1,693 in 1999-2000 and 1,508 in 2000-01 due to increased sanctions on airlines bringing in undocumented persons. In 1998-99, 926 people arrived without authority on 42 boats, compared with 157 on thirteen boats in 1997-98 (an increase of 490 percent) and in 1999-2000, 4,174 people on 75 boats, an increase of 351 percent over the previous year.

The undocumented migrants arriving by air arrive either with no travel documents or present documentation which is found to be fraudulent but which they might have used for check-in at overseas airports. While many arrive as individuals, planning their own travel, some are part of organized people trafficking organizations which have become more active across the Asian region.

Turning to people who seek to clandestinely enter Australia by boat, it needs to be pointed out that there is a history of the arrival of such boat people but that the recent increase in the numbers detected has been unprecedented. In the period from 1975 to 1989, 2059 Vietnamese boat people were intercepted as part of the refugee exodus from that region

(Viviani, 1996, 151). However, since June 1989, some 13,483 people have been detected on boats illegally attempting to enter Australia, although in the last twelve months there have been very few arrivals.

It is apparent that Australia was targeted by people smugglers who have been active in facilitating undocumented migration in other parts of the world (especially North America and Europe) for many years. The proliferation of the global international migration industry has a major illegal element which is becoming stronger over time and more widespread in its activity, so that it is becoming one of the most substantial areas of international crime. Moreover, large international crime syndicates, including many involved in the international drug trade, are becoming increasingly involved. Until recently, Australia has not been a major target of this activity but it is clear that this has now changed and that we can expect more undocumented migrant arrivals.

Australia's policy of detention of all arrivals has been the subject of much attention in 2000. To quote from DIMA (2000c):

> Australia's *Migration Act 1958* requires that all non-Australians who are unlawfully in Australia must be detained and that, unless they are granted permission to remain in Australia, they must be removed from Australia as soon as practicable. This practice is consistent with the fundamental legal principle, accepted in Australian and international law, that in terms of national sovereignty, the State determines which non-citizens are admitted or permitted to remain and the conditions under which they may be removed.
>
> The Government seeks to minimize the period of time taken to process applications made by detainees and hence the period of detention. The majority of people in immigration detention are held for a short time – in some cases as little as a few hours. However a number of factors can contribute to increased periods in detention, including court appeals and delays in the procurement of travel documents for removal.

Immigration and Regional Development

One of the distinctive features of contemporary immigrant settlement in OECD states is their concentration in a relatively small number of places, so that their spatial distribution differs significantly from that of the native population. In particular, both permanent and temporary immigrants have concentrated in the major urban centres of OECD states, especially those with substantial global linkages who could be categorized as world or global cities (Friedmann, 1986; Sassen, 1991). Australia is not an exception with Sydney (41.3 percent) housing the majority of recent arrivals (that is, at the 1996 census those who have been resident less than five years). Social networks are of crucial importance in this concentration and this tends to militate against attempts by governments to attract newly-arrived immigrants to other areas (Hugo, 1999c). There are also debates regarding the tendency for some immigrant groups to concentrate in particular areas. There is quite a bit of research evidence that such ethnic enclaves can play a positive role in cushioning the adjustment of new arrivals because they can be assisted in getting housing and work by compatriots as well as being able to use their own language in shopping and accessing services. Such endeavours have also, in some cases, been incubators of ethnic business activity. Others argue that the enclaves foster separatism and are divisive, discouraging integration to mainstream society.

Immigration and Poverty

Over the entire post-war period a debate has continued in Australia about the relative significance of two social mobility models. On the first, immigrants experience considerable upward mobility, especially between generations. While they may begin on low incomes, by dint of hard work, thriftiness and education, they have been able to improve their position. The other argument suggests that immigrants get trapped into an underclass. They experience considerable barriers to mobility in the shape of language problems, lack of education, lack of recognition of qualifications, discrimination, etc. and get trapped into low income, low well-being situations. In fact, both models apply, with there being considerable contrast between and within different birthplace groups. Existing migration policies, in fact, tend to result in a bipolar distribution of immigrants with respect to income and well-being. The increasing skill orientation of policy has meant that immigrants are well represented among the better-off segments of society. On the other hand, some groups, especially some refugee groups, experience great difficulty in entering the labour market and are strongly represented among the poor.

Student Migration

Australia has sought to attract substantial numbers of full-fee paying foreign students, especially at the tertiary level, and, particularly from the Asian region (Shu and Hawthorne, 1996). In Australia over the 1987-2000 period, the number of full-fee overseas students in Australia increased from 7,131 to 188,277 (DEETYA, 1995; DETYA, 2001). The 1997 crisis in Asia had some impact but the numbers of student visas given off-shore increased by 6 percent to 67,130 over 1998-99, by 11 percent to 74,428 in 1999-2000 and by 15.9 percent to 86,277 in 2000-01. The major sources are the USA (7,426 visas) and Asian countries, such as the PRC (8,886),

Indonesia (6,074), Malaysia (6,236) and Hong Kong (5,740). Australia has become one of the world's major destinations of international student migration. In 1998, it accounted for 8 percent of all international students in the OECD states (OECD, 2001, 32). It has become one of the top five export-earning activities in the country. A number of issues surround the migration, including:

- The extent to which it is the first step to eventual settlement, Australia has recently made it easier for students completing study in Australia remaining after graduation.
- Within Australia there is debate as to whether the increasing significance of foreign student numbers is diverting educational resources away from native students, although the evidence points to the opposite actually being the case.
- The concentration of foreign students in areas of study closely related to the New Economy (information technology, management, engineering, accountancy, etc.).

The Role of Temporary Migration

Non-permanent immigration of workers is increasing in scale. Since workers given visas are required to have employment or be in selected areas of work which lack skilled labour (e.g. information technology), the numbers involved are largely driven by labour demand. There is debate as to whether this movement is being used as a substitute for long-term development of the workforce, etc. as has been pointed out earlier. There is apparently a significant number of entrants under these schemes who later apply for permanent residence so it may be that a new form of immigrant settlement may be emerging whereby people enter the country under some form of temporary visa but become settlers. This compares to the model proposed by Castles and Miller (1993, 25) based largely on the European context but which has not previously been relevant to the Australian context. Issues arise also about the extent to which temporary migration is used by employers to keep down wages and conditions and as a way of exploiting workers. There are pressures from some employers (for example, in seasonal fruit picking areas) to extend the scheme to unskilled workers.

The Immigration Industry

Globally, the expansion of international migration has seen the proliferation of groups who play a role in encouraging and facilitating that movement. This includes groups such as immigration lawyers, recruiters, travel agents, etc. who work within the legal framework but also substantial numbers who operate outside of it. Indeed, much of the current undocumented migration could not have occurred if it wasn't for the involvement of a range of gatekeepers in facilitating the movement. These groups are part of international networks which are, in some cases, organized criminal syndicates. People trafficking has emerged as a very important and international crime in the Asia-Pacific region. There is a deal of discussion as to how this industry should be engaged, especially those involved in undocumented migration.

The Brain Drain

The issue of the net flow of highly educated and skilled immigrants from less developed to more developed countries is becoming an issue of major global significance. However, it is a much more complex issue than appears at first glance. This, of course, involves consideration of the rights of individuals as well as communities and nations. Undoubtedly, in situations where states have invested in their human capital and then see that human capital move to other states, there is an important issue. However, it has been argued that the loss of these people does not always represent a net loss to their country of origin. The reasons for this are as follows:

- In some states there is an overproduction of people with some skills so that all of the products of the education/training system cannot be productively absorbed in the labour market.
- Many such movers intend to and eventually return to their home state.
- There are substantial flows of remittances from Indians overseas which have played a major role in Indian development.
- Indians overseas have created significant opportunities for investment in India from the countries in which they have settled.
- Indians overseas have served as beachheads for the sale of Indian goods overseas.

Hence in some contexts the Indian diaspora has been a significant net asset to the state. Nevertheless, the brain drain issue is of great complexity and its full impact is yet to be established.

Conclusion

The exchange of people between India and Australia has undergone substantial changes over the post-war period. However, the understanding of the consequences of this movement for India and Australia and for the movers is little understood. There is a pressing need for this to be addressed.

In considering the future of international migration between India and Australia, it can be confidently predicted that the level and complexity of the interaction will increase over the next decade. The upsurge in settlement of Indians in Australia in recent years has led to an increase in the strength and spread of the social networks linking the two states. Moreover, the immigration industry in both countries is strengthening and will continue to both encourage and facilitate movement. It is also likely that the movement will continue to be greater from India to Australia than in the opposite direction. However, as the economies of the countries become more closely linked, it is likely that there will be some increase in movement of skilled Australians to work and live in India, as has been the case in some other Asian states. India has a large reservoir of people who readily fit the criteria currently adopted in Australia for selection of settlers, and for entering the country as temporary workers, and these flows are likely to get greater. There is a need to investigate more the implications and impacts of this movement, not only on the movers themselves but also on the communities and countries of origin and destination. Only then, will there be a sound basis for policy development which maximizes the benefits of the movement and minimizes its negative effects.

Notes

1. However, before 1974 former settlers were not classified as emigrants unless they had been in Australia for at least 12 months.
2. The current TFR in Australia is 1.75
3. Australia's official multiculturalism stance not only involves protection of rights of all migrants but also a substantial approach to involve cultural maintenance, language maintenance and other elements which encourage immigrants to maintain their heritage while also becoming full members of Australian society.

References

Australian Bureau of Statistics (ABS). *Migration Australia,* various issues, ABS, Canberra.

Australian Bureau of Statistics (ABS), 2000. *How Australia Takes a Census,* Catalogue No. 2903.0, ABS, Canberra.

Australian Bureau of Statistics (ABS), 2001. *Australian Social Trends 2001,* Catalogue No. 4102.0, ABS, Canberra.

Birrell, B., 1998. An Evaluation of Recent Changes to the Roles Governing the Entry into Australia of Skilled Business Persons and Doctors for Temporary Employment. Paper presented at Australian Population Association Ninth National Conference.

Birrell, B., Dobson, I.R., Rapson, V. and Smith, T.F., 2001. *Skilled Labour: Gains and Losses,* Centre for Population and Urban Research, Monash University.

Castles, S. and Miller, M.J., 1993. *The Age of Migration: International Population Movements in the Modern World,* Guilford Press, New York.

Commonwealth Bureau of Census and Statistics (CBCS). *Demography,* various issues, CBCS, Canberra.

Department of Education, Training and Youth Affairs (DETYA), 2001. *Overseas Student Statistics 2000,* AGPS, Canberra.

Department of Immigration and Multicultural Affairs (DIMA). *Australian Immigration Consolidated Statistics,* various issues, AGPS, Canberra.

Department of Immigration and Multicultural Affairs (DIMA). *Immigration Update,* various issues, AGPS, Canberra.

Department of Immigration, Multicultural and Indigenous Affairs (DIMA), 2000a. *Immigration Update June Quarter 2000,* AGPS, Canberra.

Department of Immigration and Multicultural Affairs (DIMA), 2000b. *Population Flows: Immigration Aspects, 2000 Edition,* AGPS, Canberra.

Department of Immigration and Multicultural Affairs (DIMA), 2000c. *Immigration Detention,* DIMA Fact Sheet 82, DIMA, Canberra.

Department of Immigration, Multicultural and Indigenous Affairs (DIMIA), 2001. *Temporary Entrants 1999-2000,* AGPS, Canberra.

Department of Immigration, Multicultural and Indigenous Affairs (DIMIA), 2002a. *Population Flows: Immigration Aspects, December 1999,* AGPS, Canberra.

Department of Immigration, Multicultural and Indigenous Affairs (DIMIA), 2002b. *Immigration Update 2001-2002,* AGPS, Canberra.

Findlay, A.M., 1990. A Migration Channels Approach to the Study of High Level Manpower Movements: A Theoretical Perspective, *International Migration,* 28(1), 15-24.

Friedmann, J., 1986. The World City Hypothesis, *Development and Change,* 17, 69-83.

Hugo, G.J., 1986. *Australia's Changing Population: Trends and Implications,* Oxford University Press, Melbourne.

Hugo, G.J., 1994. *The Economic Implications of Emigration from Australia,* AGPS, Canberra.

Hugo, G.J., 1996. *Atlas of the Australian People – 1991 Census, Western Australia,* AGPS, Canberra.

Hugo, G.J., 1999a. A New Paradigm of International Migration in Australia, *New Zealand Population Review*, 25(1-2), 1-39.

Hugo, G.J., 1999b. *Atlas of the Australian People 1996 Census: National Overview*, DIMA, Canberra.

Hugo, G.J., 1999c. *Regional Development Through Immigration? The Reality Behind the Rhetoric*, Department of the Parliamentary Library, Information and Research Services, Research Paper No. 9, 1999-2000, Department of the Parliamentary Library, Canberra.

Hugo, G.J., forthcoming. From Compassion to Compliance?: Trends in Refugee and Humanitarian Migration in Australia. Paper prepared for consideration for a Special Issue of *GeoJournal* on the Geography of Refugees.

Jones, G., 1998. 'Australian Identity', Racism and Recent Responses to Asian Immigration to Australia, pp. 249-266 in E. Laquian, A. Laquian and T. McGee (eds.), *The Silent Debate: Asian Immigration and Racism in Canada*, Institute of Asian Research, University of British Columbia, Vancouver BC.

Khoo, S.E., 1989. *Census 86: Data Quality - Ancestry*, Australian Bureau of Statistics, Canberra.

Massey, D.S., Hugo, G.J., Arango, J., Kouaouci, A., Pellegrino, A. and Taylor, J.E., 1998. *Worlds in Motion: Understanding International Migration at the End of the Millennium*, Clarendon Press, Oxford.

McKnight, T.L., 1969. *The Camel in Australia*, Melbourne University Press, Carlton, Victoria.

Organization for Economic Cooperation and Development (OECD), 2001. Student Mobility Between and Towards OECD Countries: A Comparative Analysis. Paper presented at 'International Mobility of Highly Skilled Workers: From Statistical Analysis to the Formulation of Policies', Seminar organized by OECD (DSTI/DEELSA), Paris, 11 and 12 June.

Paice, J., 1990. The 1990s – Is the Australian Census of Population and Housing Relevant? Paper presented at the Australian Population Association Conference, Melbourne, November.

Richardson, S., Robertson, F. and Ilsley, D., 2001. *The Labour Force Experience of New Migrants*, AGPS, Canberra.

Richmond, A.J., 1991. International Migration and Global Change. Paper presented at International Conference on Migration, Centre for Advanced Studies, Faculty of Arts and Social Sciences, National University of Singapore, February.

Ruddock, P., 2002. Record Temporary Entrants Contribute to Economy, *DIMA Media Release*, MPS 1/2002, 7 January.

Sassen, S., 1991. *The Global City: New York, London, Tokyo*, Princeton University Press, Princeton.

United Nations, 2000. *Replacement Migration: Is it a Solution to Declining and Aging Populations?*, United Nations, New York.

Viviani, N., 1996. *The Indochinese in Australia, 1975-1995*, Oxford University Press, Melbourne.

Waddell, C.E., 1979. *The Well-Being of Perth's Indian Residents*, in R. Johnson (ed.), Immigrants in Western Australia, University of Western Australia Press, Perth.

Young, C.M., 1988. Towards a Population Policy: Myths and Misconceptions Concerning the Demographic Effects of Immigration, *Immigration Quarterly*, 60, 220-230.

Zlotnik, H., 1987. The Concept of International Migration as Reflected in Data Collection Systems, *International Migration Review*, Special Issue, Vol. 21.

OCEAN SECURITY
ISSUES

CHAPTER 11

Indian Ocean Maritime Boundaries: Jurisdictional Dimensions and Cooperative Measures

Vivian Louis Forbes

Environmentalists attending the Earth Summit in Johannesburg, South Africa, in August 2002, to seek a global agreement to salvage the world's depleted fisheries were informed that poor management of the world's oceans is destroying ecosystems and threatening the sustainability of the fisheries. Within a couple of months, an appeal was made by the Director of the International Maritime Bureau (IMB) for the intensification of patrols by naval and police of coastal and island states following the incident in the Gulf of Aden, off the Yemeni coast, when a blast ripped through a French-flagged tanker, the m.v. *Limburg*. The ship was allegedly rammed by a small boat. A few days later, on 19 October 2002, at around 0914 (local time), a six-metre long wooden motorboat approached a large container ship in the Hugli River, West Bengal, India. A person, allegedly armed with a long knife, boarded the ship at its stern using a long hook attached to a rope. The alert crew of the ship sounded an alarm and the would-be robber (pirate) jumped overboard and escaped. The *ICC Annual Piracy Reports*, have recorded significant increases in the number of piracy attacks on the world's oceans, notably 335 attacks in 2001; 370 in 2002; and, in the first half of 2003 there were at least 150 incidents, of which 64 attacks were reported within Indonesian archipelagic waters.

Such concerns and incidents are amongst a myriad of issues – jurisdictional and political – which have produced an array of legal-political documents-international treaties – that have implications throughout the world, and particularly in the Indian Ocean Region. In this Chapter, the jurisdictional dimensions of the geopolitical issues and possible cooperative measures are discussed in the concept of delimited and undefined maritime boundaries in the Indian Ocean Region and in the varying states' practice in interpreting international law. The Chapter will provide an overview of the delimited maritime boundaries, highlight some unusual claims to extended jurisdiction, bring to the fore some of the jurisdictional issues, focus on some of the positive aspects of cooperation, and argues for the need for the regional finalization of negotiated boundaries and the declaration of legal limits to the continental shelf.

Introduction

The delimitation of maritime boundaries in the Indian Ocean Region in the timeframe 1970-1990 has demonstrated that littoral and island states of the Indian Ocean Basin had the foresight to realize the need to establish seabed jurisdiction in anticipation of the entry into force of the 1982 *Third United Nations Convention on the Law of the Sea* (the 1982 Convention). This Convention entered into force on 16 November 1994, and, by 22 August 2003, the number of states that have ratified the 1982 Convention numbered 143. An Agreement relating to the Implementation of Part XI of the 1982 Convention has been accepted in principle by 115 states and 36 have ratified an Agreement for the Implementation of the Provisions relating to Conservation and Management of Straddling Fish Stocks and Highly Migratory Fish Stocks. [Appendix 11.1, as cited in theUN Division for Ocean Affairs and Law of the Sea Web pages, 2003 – updated periodically]

A mere 28 states of the Indian Ocean Basin have ratified the 1982 Convention; 21 have noted their acceptance of the Agreement relating to Part XI of the 1982 Convention; and, only seven states have ratified the Agreement relating to the conservation and management of straddling fish stock and highly migratory fish stocks.

However, in the Indian Ocean Region, there exist a number of states that have yet to ratify the 1982 Convention, to enact legislation (or perhaps to publicize if enactment is in place) so as to implement the provisions of the 1982 Convention and other international treaties, to formulate a National Ocean Policy and to finalize negotiations on the establishment of maritime boundaries with their neighbours. There are a number of disputes over sovereignty issues and over the precise limits of the terrestrial boundary between two states. Selected maritime boundary disputes are discussed below.

The Geographical and Geopolitical Context

A decision in early 2000 by the International Hydrographic Organization to delimit a fifth world ocean from the southern portions of the Atlantic Ocean, Indian Ocean, and Pacific Ocean, shrinks the surface area of the Indian Ocean basin. Despite this truncation, the Indian Ocean remains the third-largest of the world's five oceans (after the Pacific and Atlantic Oceans, but larger than the Southern and Arctic Oceans). The new ocean – the Southern Ocean – extends from the coast of Antarctica northwards to Latitude 60° South, the delineation of which coincides with the *Antarctic Treaty* Limit.

The revised surface area of the Indian Ocean basin is about 68,560 million square kilometres. It includes the Andaman Sea, Arabian Sea, Arafura Sea, Bay of Bengal, Great Australian Bight, Gulf of Aden, Gulf of Oman, Mozambique Channel, Persian Gulf, Red Sea, Timor Sea, and Straits of Malacca and Singapore, and other interconnecting tributary water bodies. The major geographical constrictions – the choke points – include Bab el Mandeb, Strait of Hormuz, Straits of Malacca and Singapore, the southern access to the Suez Canal and the Lombok and Sunda Straits.

The ocean floor is dominated by the Mid-Indian Ocean Ridge and subdivided by the Southeast Indian Ocean Ridge, Southwest Indian Ocean Ridge, and Ninety-East Ridge. The lowest point is the Java Trench at 7,258 metres, which lies south of Java Island in the northeast sector of the basin. Further east is the Timor Trough with a depth of over 3,500 metres. Comparatively wide continental margins are found to the north and west of the Australian coast and off the Indian sub-continent, and only a comparatively narrow shelf along the East African littoral.

The Indian Ocean provides major sea routes connecting the Middle East, Africa, and East Asia with Europe and the Americas. It carries a particularly heavy traffic of petroleum and petroleum products from the oilfields of the Persian Gulf and Indonesia. Its fish are of great and growing importance to the population of the littoral and island states for domestic consumption and export. Fishing fleets from Russia, Japan, South Korea, and Taiwan also harvest the Indian Ocean, mainly for shrimp and tuna.

The natural resources – marine, biotic and mineral – include hydrocarbon reserves, fish, shrimp, sand and gravel aggregates, placer deposits, hot brines and polymetallic nodules. The current environmental issues in the region are endangered marine species that include the dugong, seals, turtles, and whales; oil pollution in the Arabian Sea, Persian Gulf, and Red Sea; and coastal erosion, mangrove destruction and associated habitat loss and coral reef bleaching. Large reserves of hydrocarbons are being tapped in the offshore sedimentary basin areas in the Persian Gulf, northern Red Sea, India, and Western Australia, and the potential hydrocarbon reserves of the East African littoral. An estimated 40 per cent of the world's offshore oil production comes from the Indian Ocean region. Beach sands rich in heavy minerals and offshore placer deposits are actively exploited by bordering states, particularly India, South Africa, Indonesia, Sri Lanka, and Thailand.

The major geopolitical concerns in the Indian Ocean region are terrorism, religious divides, acts of piracy, the threat of war in the Middle East and the sub-continent, the demand for regional autonomy in some countries of the Basin and the recognition of indigenous rights to territory and resources – onshore and offshore. The rights to, and obligations on, coastal and island states in establishing maritime jurisdictional limits are embodied in international law.

Maritime Jurisdictional Limits

The 1982 Convention replaced the four 1958 *Geneva Conventions* (The Second Law of the Sea Convention) in 1994 and is now crystallized in international law. In addition to the *Internal Waters*, the *Territorial Sea* and the *Contiguous Zone*, five additional jurisdictional zones are now recognised. These are: the *Continental Shelf*, the *Exclusive Economic Zone* (EEZ), the *Exclusive Fishing Zone, Archipelagic Waters* and the *Area*. Their existence permits coastal states to claim rights over the adjacent seas for a variety of purposes and over varying but specified distances from an established datum (Bernaerts, 1988; Forbes 1995; Prescott, 1985).

The starting point for measuring the breadth of the various jurisdictional zones extending from the coast seaward is the *Territorial Sea Baseline*. Naturally, until a coastal state establishes its baseline, it is generally unable to define and delineate its maritime boundary, although there are instances where maritime boundaries, however defined, have been finalized even though the state's baselines have not been proclaimed, based on current knowledge - for example, India, Malaysia and Tanzania.

Article 5 of the 1982 Convention states that "the normal baseline for measuring the maritime zones is the low-water mark along the coast which is marked on large-scale (nautical) charts officially recognized by the coastal state". The datum for the low-water mark or low-tide elevation (LTE) has been defined in different ways (Article 13).

Different countries thus use various data based on a range of low-water marks, such as the Lowest Astronomical Tide (LAT), Indian Springs Low Water (ISLW), Mean Low-Water Springs (MLWS), Mean Low-Water Neaps (MLWN) and Mean Lower Low-Water (MLLW). The MLWS was advocated by delegates at an international conference at The

Hague in 1930. The MLWS, also called springs low-water, is defined in the *Hydrographic Dictionary* as being "the average height of the low waters of Spring Tides", measuring successive low-waters during the period when the moon's declination is at its maximum of 23°30'. There have also been moves to use the LAT, a level which can be predicted to occur under average meteorological conditions, and under any combination of astronomical conditions.

In the case of islands comprising parts of atolls, or of islands having fringing reefs, the baseline is determined by the LAT (Article 6). Where a coastline is deeply indented or there is a fringe of islands along the coast, straight baselines joining appropriate points on the coastline may be used in drawing the baseline from which the breadth of the maritime zones is measured (Article 7). These lines must not deviate to any appreciable extent from the general direction of the coast. The sea areas lying within the lines must be sufficiently closely linked to the land to be subject to the regime of internal waters.

Article 7:4 of the 1982 Convention stipulates that straight baselines may not be drawn to and from LTE, unless they pass under structures (lighthouses or similar installations) which are permanently above sea level. An exception exists in instances where the drawing of the baselines to and from such elevations has received general international recognition. On the other hand, Article 13 provides that, if the LTE lies within the territorial sea claimed from the coast of the mainland or an island, the low-water line on that elevation may be used as a point for a baseline system. However, if the LTE, which is located on a naturally-formed area of land and which is exposed at low-tide level, lies at a distance that exceeds the breadth of the territorial sea from the mainland or an island, it may not be used to generate a further territorial sea of its own. A coastal state may not apply a system of straight baselines in order to cut off the territorial sea of another state from the high seas or from an exclusive economic zone.

Coastal states may determine baselines by any method, or a combination of methods, as provided for in Articles 5 to14, and these baselines must be shown on charts of a scale or scales adequate for ascertaining their position. Alternatively, a list of geographical coordinates, specifying each geodetic datum, may be substituted. Either of these types of information must be given due publicity and a copy of each chart or list must be deposited with the Secretary-General of the United Nations (Article 16).

Archipelagic states have a special set of rules (Article 47) for the establishment of baselines. For the purpose of the 1982 Convention, such a state is defined as composed "wholly by one or more archipelagos and may include other islands" (Article 46:a). An archipelagic state is permitted to draw straight archipelagic baselines joining the outermost points of the outermost islands and drying reefs of the archipelago, provided that within such baselines are included the main islands and an area in which the ratio of the area of the water to the area of the land, including atolls, is between 1:1 and 9:1.

The special formula for drawing baselines to close a bay was first prescribed in the Geneva Convention of 1958, and was retained in an amended form in the 1982 Convention. An indentation is not regarded as a bay unless its area is as large as, or larger than, that of the semi-circle whose diameter is a line drawn across the mouth of that indentation. On the other hand, in this context a river mouth is treated in the same way as the entrance to a bay (Article 9). A bay-closing line may be drawn between the low-water marks of the natural entrance of a bay provided that the line does not exceed 24 nm in length. The waters enclosed by this line are considered as being internal waters of the coastal state.

Except as provided in Part IV (Archipelagic States) of the 1982 Convention the waters on the landward side of the proclaimed baseline of a coastal state form a part of the Internal Waters of that state (Article 8: i). Such waters are created when baselines other than the natural low-water mark are selected. It must be noted, however, that where a straight baseline has the effect of enclosing internal waters areas which had not previously been considered as such, the right of innocent passage exists in those newly created internal waters.

Internal Waters and the Territorial Sea

The Territorial Sea is that part of the open sea over which a coastal state may claim jurisdiction for its own protection. Although traditionally one league (3.5 statute miles), the breadth of the territorial sea has been a matter of controversy for many decades, with seaward limits varying from three to 200 nm. Article 3 of the 1982 Convention reserved for every coastal and island state the right to establish the breadth of its territorial sea up to a limit not exceeding 12 nautical miles, measured from the baselines determined in accordance with the Convention.

In the Indian Ocean region, in 2003, for example, while three states still claimed a territorial sea of three-nautical mile breadth, 34 states claim a width of 12-nautical miles, and Somalia claims a 200-nautical mile zone (See Appendix 11.2). The United Kingdom claims a three-nautical mile zone around the British Indian Ocean Territories (BIOT) (Forbes, 1992:16). The Territorial Sea, the airspace above it, its bed and substratum remain under the sovereignty of the coastal state (Article 2), which, however, has an obligation to accord the right of innocent passage to foreign shipping (Article 17). Provision for the passage of commercial and military ships and aircraft through and over such straits (Articles 37 - 44), along similar lines to the right of freedom of navigation on the high seas, was a concession that the principal maritime powers secured during the negotiations.

The Contiguous Zone

Article 24 of the 1958 *Geneva Convention on the Territorial Sea and Contiguous Zone* entitled a coastal state to claim as a Contiguous Zone waters extending out to 12 nautical miles from the baseline; to operate in such waters in order to prevent infringement of its customs, fiscal, immigration or sanitary regulations within its territory or territorial sea; and to punish infringement of the above regulations committed within its territory or territorial sea. In other words, the coastal state was given partial jurisdiction over a limited area of what were conceived as international waters.

Article 33 of the 1982 Convention contains a similar provision, except that the width of such waters has been extended to 24 nm measured from the baseline, and the words, "in a zone contiguous to its territorial sea, described as the contiguous zone" have been used, to indicate that this zone is no longer a part of the high seas. In any case, the width of the Contiguous Zone must not exceed 12 nautical miles. Thus, should a state claim a territorial sea width of only six nautical miles, its permissible CZ breadth is limited to the 12 nautical mile limit and not 18 nautical miles as some states would suggest.

Within the Contiguous Zone of a coastal state, aliens have full navigation rights, provided they do no not infringe regulations relating to the Territorial Sea regime. By contrast, overflight rights may be restricted. Aliens also possess the rights to conduct scientific research in the water column and to fish for all but sedentary species, if the coastal state has not proclaimed a fishing zone or an exclusive economic zone extending beyond the seaward limit of its territorial sea. They may also lay submarine cables within this zone, but do not possess any mining rights. A coastal state may impose environmental legislation on the contiguous zone provided it is overlapped by an EEZ or underlain by a Continental Shelf. In the first instance, the coastal state has authority over the seabed and the water column; in the second, it can only legislate for the seabed (Article 194).

The Contiguous Zone thus enjoys specific legal status only as long as the coastal state has not proclaimed an EEZ (Article 57) exceeding the outer limits of the contiguous zone.

The Continental Shelf

According to the 1982 Convention, the regimes of the Territorial Sea and the Contiguous Zone take precedence over the regime of the Continental Shelf in that part of the seabed that underlies the water columns of the territorial sea and contiguous zone. Geomorphologically, the continental shelf is defined as the zone around a continent extending from the low-water line to a depth at which the ocean bottom slopes markedly downward, and away from the continent. Conventionally, the edge of the shelf was taken to lie at 200 metres, or 100 fathoms. A more accurate average depth is about 130 metres. Continental shelves underlie about eight percent of the total ocean area and coastal states have increasingly tended to regard them as a natural continuation of their land area, and, therefore, as part of their national territory.

Whereas the coastal state may exercise its rights to explore and exploit its natural resources over the continental shelf (Article 77: 1), these rights "do not affect the legal status of the superjacent waters or of the air space above those waters" (Article 78: 1). Furthermore, the rights of the coastal state over its continental shelf should not infringe or interfere with safe navigation and freedom of the seas. In addition, within this zone, aliens have full navigation rights, provided they observe the rules and safety zones designated by the coastal state. They also possess the rights to lay submarine cables; conduct research in the water column, but with consent from the coastal state, if the research is of the seabed; fish in the water column, but not catch sedentary species (Article 77: 4). Coastal and island states have complete authority to legislate for the protection of the seabed environment (Articles 194 and 200) provided that such regulations do not unduly interfere with the rights and duties of aliens. The Convention also permits aliens to conduct research in those areas of the shelf which are more than 200 nm from the baseline of the coastal state and which have not been designated by that state as areas within which exploration ". . . will occur within a reasonable period of time" (Article 246: 6).

The Exclusive Economic Zone and Fishing Zones

Although the Truman doctrine of the Continental Shelf provided for the United States dominion over the land beneath the sea, it was not sufficiently specific about the exploitation of living non-sedentary marine resources. Concern over this issue during the 1960s and 1970s gave rise to the concept of an Exclusive Economic Zone (EEZ), which became a specific legal regime created by the 1982 Convention. Initially proposed by Kenya at Lagos in 1972, the concept of the EEZ had its origins in the claims of Chile, Peru, Costa Rica and El Salvador to 200 nautical mile fishing zones in their respective adjacent waters in the late 1940s (Attard, 1989). Article 57 of the Convention gives a coastal state the right to establish an EEZ extending up to 200 nm seawards from the baseline and requires the state to provide details of its outer limits. Within the zone, the state has preferential fishing rights and exclusive control over the exploitation of mineral resources. Archipelagic states and many coastal states would gain substantial advantages in claiming the full width of such a zone, which is defined as an area beyond and adjacent to the territorial sea. A number of exhaustive studies of the regime of the EEZ have been compiled independently by Attard (1989), Kwaitkowska (1989) and Smith (1986).

If a conflict arises between the interests of the coastal state and any other state or states, Article 59 provides for a resolution on ". . . the basis of equity and in the light of all the relevant circumstances . . ." Article 60 provides that, within the EEZ, the coastal state has exclusive rights to construct, operate and maintain artificial islands, installations and other structures. Article 61 of the 1982 Convention requires the coastal state to implement conservation and management measures to ensure the maintenance of fish population levels which can produce the maximum sustainable yield as qualified by relevant environmental and economic factors. The coastal state, under Article 62, is also required to ". . . promote the objective of optimum utilization of the living resources in the exclusive economic zone. . .". The Article further notes that ". . .where the coastal state does not have the capacity to harvest the entire allowable catch, it shall, through agreements or other arrangements, give other states access to the surplus of the allowable catch, having particular regard to the provisions of Articles 69 [Rights of landlocked states] and 70 [Rights of geographically disadvantaged states], especially in relation to the developing states mentioned therein". The 1982 Convention, however, does not attempt to elaborate on the meaning of the objective "optimum utilization", the definition of which is likely to change with technological development and is left to the discretion of the coastal state.

Provisions for the enforcement of a coastal state's laws and regulations within the EEZ are contained in Article 73, which stipulates that the penalties for violation may not "include imprisonment, in the absence of agreements to the contrary by the states concerned", or other forms of corporal punishment, but that offenders may be required to post a reasonable bond or other security, to obtain release after being arrested. Perhaps the most vexing question concerning the EEZ is its delimitation between states with opposite or adjacent coasts. In order to achieve an equitable solution, Article 74 proposes that the delimitation "be effected by agreement on the basis of international law", and cites Article 38 of the Statute of the International Court of Justice, in order to achieve such a solution. A provisional arrangement of a practical nature, without prejudice to the final delimitation, would be preferable to failure to proclaim the outer limits of an EEZ. However, where there is an agreement in force between the coastal and or island states, the delimitation issue should be determined in accordance with the provisions of the existing agreements (Article 74:4).

The expanse of each state's EEZ varies with the length and configuration of its coastline. India, with its long coastline and numerous island territories, could claim a zone of about 588,000 sq nm, whereas Kenya, with the shortest coastline of Africa's Indian Ocean littoral states, could claim a zone of about 34,000 sq nm. The island state of Singapore, being zone-locked by its coastal neighbours of Indonesia and Malaysia, has, not surprisingly, consistently opposed the establishment of the EEZ, for it could only claim a very modest area of the seas beyond its territorial jurisdiction. In the Indian Ocean basin, 27 states have confirmed that they claim the full 200-nautical mile limit as their EEZ; nine States have not specified, or perhaps have not publicized their legislation; and two States, namely, Jordan and Singapore, due to their 'geographically disadvantaged positions' are limited to a little more space beyond their territorial sea limits, that is three nautical miles.

Beyond National Jurisdiction: the High Seas and the 'Area'

Beyond the coastal and island states' national maritime jurisdiction, that is, outside the limits of the claimed EEZ or legal continental shelf, whichever is further seaward from the landmass, the resources and the maritime space are reserved for the 'Common Heritage of Mankind'. In the legal sense, this covers the deep seabed and the ocean floor and their substratum which lie beyond the limits of national jurisdiction.

A declaration in the 1958 *Convention on the High Seas* defined the High Seas as that part of the sea which was not included in the internal waters or territorial seas of states. Article 86 of the 1982 Convention amended the definition to include only that part of the sea which is not included in the internal waters, the territorial sea or the exclusive economic zone of a state or in the archipelagic waters of an archipelagic state.

The high seas are open to all states, whether coastal or landlocked, and should be used only for peaceful purposes. According to Article 87, the five freedoms of the high seas are those of navigation; of overflight; of the laying of submarine cables and pipelines; of construction of artificial islands and other installations; and, of fishing and carrying out scientific research. With reference to conducting scientific research, however, this privilege does not include such research on the continental shelf of a coastal state which underlies the high seas (Articles 87 [f] and 257). Nearly two-thirds of the surface of the Indian Ocean falls under the regime of the high seas. No high seas remain in the Persian Gulf, Andaman Sea, Red Sea, the southern sector of the South China Sea or the Timor Sea, although the waters beyond the territorial seas are, by international customary law, considered as 'international water'.

Beneath the high seas is the *Area*, defined by Article 136 of the 1982 Convention as comprising "the seabed and ocean floor, the subsoil thereof, beyond the limits of national jurisdiction, (which) as well as the resources of the Area remain the common heritage of mankind". Its legal status is proclaimed in Article 137, under which no state may "claim or exercise sovereignty or sovereign rights over any part of the Area or its resources". If, in due course, the combined surface area of Indian Ocean claimed as EEZ is equivalent to one-third of the total surface area, then the remaining two-thirds will, with those parts of the Area located elsewhere, be "vested in mankind as a whole" and will, in accordance with Article 157, be managed by the International Seabed Authority (ISA). This body is located in Kingston, Jamaica, with regional offices set up to carry out its functions.

The resources of the Area – which include all solid, liquid or gaseous mineral resources, including poly-metallic nodules – are not subject to alienation. However, when minerals are recovered from the Area, they may, in accordance with the procedures of the Authority, be alienated. A royalty will be imposed for the privilege of extracting minerals from the seabed of the Area. Other activities in the Area, which must pay due regard to the protection of human life (Article 146) the protection of the marine environment (Article 145), and the protection and preservation of archaeological and historical objects (Article 149), are restricted to marine scientific research, exclusively for peaceful purposes and for the benefit of all of humankind.

States are urged to cooperate with the Authority to promote the transfer of technological and scientific knowledge, especially to developing states and to the Enterprise (Article 170) which will be an organ of the Authority. The Enterprise is organized with elements of a private company comprising a Director-General and a 15-member governing board. It is to operate in accordance with sound commercial principles and to carry out or oversee the transporting, processing and marketing of minerals recovered from the area (Annex IV: Statute of the Enterprise, Articles 1-13 of the 1982 Convention).

Objections Raised and Statements Made by IOR States

Article 310 of the 1982 Convention permits states and entities to make declarations or statements regarding its application at the time of signing, ratifying or acceding to the 1982 Convention, which do not purport to exclude or modify the legal effect of the provisions of the 1982 Convention, by stating:

Article 309 does not preclude a state, when signing, ratifying or acceding to this Convention, from making declarations or statements, however phrased or named, with a view, *inter alia*, to the harmonization of its laws and regulations with the provisions of this Convention, provided that such declarations or statements do not purport to exclude or to modify the legal effect of the provisions of this Convention in their application to that state.

Article 287, paragraph 1, makes provision for States and entities, when signing, ratifying or acceding to the 1982 Convention, or at any time thereafter, may make declarations specifying the forums for the settlement of disputes which they accept. It notes:

When signing, ratifying or acceding to this Convention or at any time thereafter, a State shall be free to choose, by means of a written declaration, one or more of the following means for the settlement of disputes concerning the interpretation or application of this Convention:

a) the International Tribunal for the Law of the Sea established in accordance with Annex VI;
b) the International Court of Justice;
c) an arbitral tribunal constituted in accordance with Annex VII;
d) a special arbitral tribunal constituted in accordance with Annex VIII for one or more of the categories of disputes specified therein.

In addition, Article 298, paragraph 1, allows States and entities to declare that they exclude the application of the compulsory binding procedures for the settlement of disputes under the 1982 Convention in respect of certain specified categories of disputes (UN, Division for Ocean Affairs - web).

Upon signature, on 10 December 1982, thirty-three states had made Declarations or Statements relating to certain aspects of the 1982 Convention. Five Indian Ocean states, namely, Iran, Iraq, Oman, Qatar and Sudan had lodged Declarations. Their concerns ranged from issues regarding passage through straits; the exercise of right of innocent passage of warships through the territorial sea of other states; authorization of permission to transit territorial seas; notification requirement in respect warships, of nuclear-powered ships or ships carrying nuclear or other inherently dangerous or noxious substances; the rights of states with special geographical disadvantages; full acceptance or complete rejection of the arbitral procedure; disputes concerning the interpretation or application of certain Articles; and, reserving the right to make at the appropriate time declarations concerning the settlement of disputes. In the case of Iraq, Oman and Qatar their acceptance of the document, they each noted, in no way implies their recognition of Israel nor any relationship with it (Iraq's Declaration of 10 December 1982).

By 13 October 2003, of the 52 states which had lodged Declarations or Statements at the moment of depositing their instruments of ratification or accession to the 1982 Convention, these 12 states – namely, Bangladesh, Egypt, India, Kuwait, Malaysia, Oman, Pakistan, Saudi Arabia, South Africa, the United Kingdom, Tanzania and Yemen – represented the Indian Ocean region. Each of their concerns mirrored those alluded to above.

Australia (22 March 2002) and the United Kingdom (12 January 1998 and 7 April 2003) lodged Declarations long after they deposited their instruments of ratification. The Government of Australia declared that it chooses the following means for the settlement of disputes concerning the interpretation or application of the Convention, without specifying that one has precedence over the other: (a) The International Tribunal for the Law of the Sea established in accordance with Annex VI of the Convention; and (b) The International Court of Justice.

Yemen, upon ratification (21 July 1987), noted that: 1) The People's Democratic Republic of Yemen will give precedence to its national laws in force which require prior permission for the entry or transit of foreign warships or of submarines or ships operated by nuclear power or carrying radioactive materials; and 2) With regard to the delimitation of the maritime borders between the People's Democratic Republic of Yemen and any state having coasts opposite or adjacent to it, the median line basically adopted, shall be drawn in a way such that every point of it is equidistant from the nearest points on the baselines from which the breadth of the territorial sea of any State is measured. This shall be applicable to the maritime borders of the mainland territory of the People's Democratic Republic of Yemen and also of its islands.

Malaysia noted (on 14 October 1996) that ". . . any objects of an archaeological and historical nature found within the maritime area over which it (Malaysia) exerts sovereignty or jurisdiction shall not 'be removed, without its prior notification and consent".

An overview of the delimited international political maritime boundaries – a collective term that encompasses territorial sea boundaries, seabed boundary, water column boundary, continental shelf boundaries and EEZ boundaries – that are shared in a sub-regional context, is presented below beginning in the vicinity off the East African littoral. There are at least 47 maritime boundaries that have been settled in the

Indian Ocean basin by October 2003 (Charney and Alexander, 1993; Charney and Smith, 2002; Forbes, 1995 and 2001; Prescott, 1985; US Department of State).

Delimited Maritime Boundaries

Two maritime boundaries have been established by agreement in the western margin of the Indian Ocean Basin. The boundary between Kenya and Tanzania is in force; the boundary between Mozambique and Tanzania has been negotiated, but is pending ratification, as far as can be ascertained. The Seychelles Government has defined geographical coordinates to delineate the extent of its Exclusive Economic Zone. The proclamation was based on a unilateral decision. Apparently, no discussion took place with its neighbours prior to the announcement being made. For the most part, the boundary is formed by the locus of a point 200 nm from the nearest landmass of the Republic of Seychelles except between Points 13 to 18 and 19 to 29 where the boundary is formed by the median line between the Republic and its neighbours.

There are three negotiated boundaries in the Red Sea: 1) the Common Zone of Exploitation established by Saudi Arabia and Sudan whose is the 1,000-metre isobath in that section of the Red Sea adjacent to the coasts of the two countries; 2) an arbitrated boundary between Eritrea and Yemen; and, 3) a negotiated boundary between Saudi Arabia and Yemen.

In the Persian Gulf region, maritime boundaries have been delimited for the northern and central portions of the Persian Gulf, in the Strait of Hormuz and the northern waters of the Gulf of Oman. About 11 sets of boundaries have been defined. Some were negotiated in agreements between the parties concerned; that between Iraq and Kuwait was defined by the parties, but under supervision of a Committee under the auspices of the United Nations; and another, namely between Bahrain and Qatar, through a decision made by the International Court of Justice. In the northern sector of Arabian Sea, Oman and Pakistan have delimited a boundary, and further south, in the Laccadive Sea, boundaries have been delineated between India, the Maldives and Sri Lanka.

Maritime boundaries have been determined by bilateral and trilateral agreements in the Andaman Sea, southwest corner of the Bay of Bengal, Palk Bay, Gulf of Manna, Johor Straits, the Malacca Strait and the southern sector of the South China Sea. In the Straits of Singapore a boundary between Indonesia and Singapore was delimited in 1975.

In the vicinity of the Indonesian Archipelago and adjacent regions, Indonesia has gained vast areas of sea space which now fall within its archipelagic waters regime by virtue of its declaration of archipelagic status and the adoption of the straight baseline system, which was defined in 1960 and subsequently modified, connecting the outermost islands of the archipelago. It is apparent from the geographical extent of the archipelago that Indonesia was keen and indeed took the initiative in many instances to bring about negotiations on determining continental shelf boundaries with its neighbours in the adjacent seas. On 17 December 2002, in a judgement handed down by the International Court of Justice, it was recognized that Pulaus (Islands) Ligitan and Sipidan fell under the sovereignty of Malaysia, and not Indonesia. Such a decision has implications for the future alignment of the maritime boundary between the two states.

A Treaty to delimit maritime boundaries between Christmas and Java Islands and a westward extension of Point A25 of a previously agreed seabed boundary in the Timor Sea was signed in March 1997 by the Governments of Australia and Indonesia but had not been ratified by 30 October 2003. This agreement established a seabed regime for one country in the Arafura and Timor Seas and a water column regime for the other country. Non-ratification of the Treaty has created one large 'grey area' in the region. In May 2000, Australia and East Timor signed a Treaty whereby a Joint Petroleum Development Area replaced the previous Zone of Cooperation, which was established in 1989 by Australia and Indonesia.

Australia and Papua New Guinea defined a suite of jurisdictional lines for resource allocations in Torres Strait and into the Coral Sea and created the Torres Strait Protected Zone.

Maldives, Mauritius, Seychelles and South Africa have each defined, by geographical coordinates, the outer limits to their claims for the exclusive rights to explore and exploit the resources of the sea. The claim by Mauritius is intriguing. It comprises two parts. The western part, consisting of 40 turning points, encompasses Tromelin Isle. The delineated zone incorporates the common agreed boundary with France (Réunion Island). The eastern part of the claim, defined by 21 turning points labelled A to U, encompasses the islands of the Chagos Archipelago. Part of the claim overlaps the Maldives' declared Territorial Sea and Fishing Zone Rectangle. Negotiated boundaries exist between India and Maldives and India and Sri Lanka.

The complexity of the issue over sovereignty of the Chagos Archipelago is highlighted by the decision of the Administration of the British Indian Ocean Territory (BIOT) to declare a Fishing and Conservation Management Zone (FCMZ) of 200-nm radius around the Chagos Archipelago (Forbes, 1992, 12-14). The zone overlays the earlier Mauritian claim. Maritime boundaries have, as yet, to be determined between Madagascar and the neighbouring states of Seychelles, Mauritius, Comoros and France.

In the vicinity of Some Islands in the Southern Indian Ocean a continental shelf boundary between Kerguelen and Heard and McDonald Islands has been determined. It consists of eight turning points.

Potential Maritime Boundaries

To complete the political carve-up of the maritime space within national jurisdiction, some 32 potential boundary agreements will need to be drafted. There are at least nine potential maritime boundaries to be settled in the south-western sector of the Indian Ocean basin, and they include the boundaries between: Kenya and Somalia, Mozambique and South Africa, Mozambique and France, Mozambique and Comoros, Comoros and Tanzania, Comoros and France, Comoros and Seychelles, Seychelles and France, France and Madagascar and France and Mauritius when its sovereignty claim over Ile Tromelin has been settled. The complexities of the negotiations rest on the issue of sovereignty of some small but politically significant islands in the Mozambique Channel that are administered by France.

In the Red Sea, Gulfs of Aden and Aqaba boundary agreements will be required between: Somalia and Djibouti; Somalia and Yemen; Yemen and Oman; Djibouti and Yemen; Djibouti and Eritrea; Eritrea and Sudan; Eritrea and Saudi Arabia; Egypt and Saudi Arabia; Egypt and Israel; Egypt and Jordan; Jordan and Israel; and, Jordan and Saudi Arabia.

Boundaries in the process of being negotiated or to be determined are those between Kuwait and Iraq, and Iraq and Iran at the head of the Gulf. In the south-eastern sector of the Gulf, Iran and individual Emirates of the United Arab Emirates have yet to determine their common continental shelf boundary. Delays in negotiations are centred on the issue of sovereignty of some small islands, in particular, Abu Musa, that lie between the coastlines of the two states. The insular geography of the region favours Iran in boundary determination if all the islands it claims were to be considered as points of measurement.

Potential maritime boundaries for India include those with Bangladesh in the Bay of Bengal and with Pakistan in the Arabian Sea. Bangladesh has defined its territorial sea straight baseline system, which is most unusual in that the basepoints are not located on land. Instead, the points lie along the 18-metre isobath. From this baseline system, Bangladesh has delineated the boundaries of its claimed economic zone. Although this irregularly shaped zone has its merits, it remains to be seen whether the two adjacent states will recognize the claim. Failure to do so will mean that there remain three potential continental shelf boundaries to be negotiated. A boundary between Bangladesh and Burma needs to be defined.

Within the context of the defined limits of this study, there remain four sets of boundaries to be negotiated, namely between East Timor and Indonesia; Indonesia, Malaysia and Singapore at the eastern and western approaches to the Straits of Singapore; and between Malaysia and Singapore after the International Court of Justice has handed down its decision regarding sovereignty over the 'rock' Pedro Branca by one party and Pulau Batu Puteh by the other. Upon this rock, a lighthouse called Horsburgh Lighthouse, which is administered and maintained by the Marine and Port Authority of Singapore, is used as a vital aid to navigation by all ships plying the busy sea lanes.

Specific Maritime Jurisdictional Issues

Apart from the issues raised in the Declarations and Statements issued by the states and those alluded to above, there are a number of specific maritime jurisdictional issues that require immediate attention to bring about settlement or finalization of maritime boundary delimitation within the Indian Ocean Region. The Indian Ocean offers a perfect arena for boundary control if and when the maritime spaces in the adjacent seas of the states are politically divided. An examination of some issues begins with the India and Pakistan and Bangladesh and India stalemates over the delimitation of maritime boundaries, followed by an analysis of illegal fishing and finally a brief discussion of marine terrorism and acts of piracy.

India and Pakistan

The delimitation of any maritime boundary is highly complex or extremely simple depending on how one wants to look at it. On present circumstances, it is still possible that the Governments of India and Pakistan can negotiate a maritime boundary in the north-eastern sector of the Arabian Sea. However, India's 'everything-but-Kashmir' and Pakistan's 'nothing-except-Kashmir' steadfast stances are unlikely to produce any results for some time to come. However, the two governments should not allow the terrestrial boundary problem to impinge upon the maritime boundary determination. The northern waters of the Arabian Sea and its resources offer considerable scope for cooperation and development between the two countries.

In the case of the maritime boundary, the entire problem hangs on whether the boundary should be drawn from the southern shore of Sir Creek or from a reference point mid-stream, or from the northern shore of the Creek. The area involved is a few hundred square nautical miles of maritime space. With a little diplomacy by both sides, a resolution could be easily found, if there is the spirit of cooperation. Both sides continue to produce historical records, search for boundary pillars and debate whether a piece of land is new or old. The dispute resembles one of those feuds over a few hectares of land whilst there is vast maritime space to be explored and resources to harvested and exploited.

Bangladesh and India

On 1 October 2003, the Parliamentary Standing Committee on Defence of Bangladesh requested the Bangladesh Navy to ensure Bangladesh's sovereignty over South Talpatty Island. The Chair of the Committee noted that the island was an integral part of Bangladesh. The ownership of the island has been in dispute between Bangladesh and India since it emerged in the estuary of the border river, Hariabhanga, after the devastating cyclone that swept through Bangladesh in November, 1970. The island is located about two nautical miles south of the Hariabhanga River that politically divides Bangladesh and India. The approximate geographical coordinates are Latitude 21° 36.0' North and Longitude 89° 09.1' East. It is believed to be of U-shaped formation with the eastern arm elongated towards the north, and it had an approximate area at low tide of about one square nautical mile in 1978. It may have further grown since then.

Both countries claim the island as forming part of their territories. India calls it 'New Moore Island or Purbasha' whereas the name 'South Talpatty Island' in Bangladesh is derived because of its proximity to the Talpatty land area of the Sunderbans of Bangladesh. The island has been an uninhabited, although fishers from the Bangladesh mainland were often sighted seasonally during the dry season in the late-1970s.

When Bangladesh became an independent state in 1971, its government was engaged in the challenging task of reconstruction and rehabilitation of the war-ravaged country. In no time during this period did India bring Bangladesh's specific attention to the island. Bangladesh was reportedly told of the ramifications of the presence of this island in 1974 on the demarcation of the sea boundary in the Bay of Bengal when the Indian delegation visited the capital, Dhaka. Prior to the discussion of the maritime boundary in the Bay of Bengal, it was realized that, first, the question of the border of the Hariabhanga River had to be determined. Ordinarily, in the case of a navigable river, under international law, the boundary line runs through the middle of the deepest navigable channel (akin to the thalweg principle), unless otherwise agreed between the parties.

During the discussions, Bangladesh claimed sovereignty over the South Talpatty Island on the grounds that: (a) the flows of the border river were to the west of the island, and a satellite photograph confirmed the western orientation of the flow, and, (b) the island was a natural prolongation of Bangladesh territory. The satellite imagery depicted that the main channel of the Hariabhanga River turned a little right near the mouth of the river and entered the Bay of Bengal keeping South Talpatty to its left. In other words, the Hariabhanga River flows to the west of South Talpatty Island.

This was also clear from the flow of the suspended sediments entering the Bay of Bengal. India disputed Bangladesh's position, as it claimed that the flows of the border river lay to the east of the island, not west, as claimed by Bangladesh. The key point of the dispute over the sovereignty of the island rests on whether the main flow of the Hariabhanga River lies to the west or east of the island. It is pertinent to note that two flows – one from the border river and the other from Bangladesh inland river, Raimangal – flow in the estuary. Bangladesh argued that India confused the flow of the Raimangal River with that of the border river, Hariabhanga. There were many discussions at the official level over two decades to resolve the dispute, but they remained inconclusive. In April 1979, when the Indian Prime Minister, Morarji Desai, visited Bangladesh, the President of Bangladesh Ziaur Rahman took up the matter with him. In the interests of good neighbourly relations, to resolve the dispute, Bangladesh proposed a survey by a joint Indo-Bangladesh team to locate exactly the main flow of the border river around the island. The commitment was reportedly confirmed by the Indian Prime Minister when the Deputy Prime Minister of Bangladesh called on him in New Delhi in the second week of May 1979. Furthermore, in August 1980, during the visit of India's Foreign Minister to Dhaka, a Joint Press Statement was issued on August 18, 1980 and paragraph 9 of the statement read as follows:

The question of the newly-emerged island (New Moore/South Talpatty/Purbasha) at the estuary of the border river, Hariabhanga, was also discussed. The two sides agreed that, after study of the additional information exchanged between the two governments, further discussion would take place with a view to settling it peacefully at an early date.

Exchange of information in the above joint statement was intended to include the result of the joint survey to ascertain the physical location of the main flow of the border river, Hariabhanga. While bilateral discussions were pending to resolve the dispute, on 9 May 1981, India sent its naval ship INS *Sandhayak* with one helicopter and some military personnel to the island. Some huts, tents, one aerial mast and one pole bearing the Indian flag were seen erected there. Bangladesh was taken by surprise at India's claim to the island. On 11 May 1981, Bangladesh lodged a strong protest against such unwarranted, unilateral and illegal action of India which it noted was in breach of the agreements reached at the highest political level. On 16 May 1981, Bangladesh urged India to withdraw the Indian navy ship from Bangladeshi waters and to remove men and materials including the flag from the disputed island. India in reply reportedly stated that the intention of sending the ship was to collect additional hydrological information on the island. Later, India agreed to withdraw its ship and men and materials from the disputed island so that it would remain "no man's land" until it was settled peacefully. The Indian Border Security Force (BSF), which established a base on the island which was regularly visited by Indian naval ships, is cause for great concern to Bangladesh. If this is true, India is perceived as showing military might to claim sovereignty on the island, thus ignoring the legitimate claim of Bangladesh. In accordance with international law and agreements between the two sides, the sovereignty of the island needs to be settled between the two states. Bangladesh requires that India honour its commitment in the spirit in which it was made and in the interests of good neighbourly relations. Cooperation based on mutual trust provides an apt model, as evident in the following case.

Hydrocarbon Potential and Boundary Issues

Kuwait and Saudi Arabia intend to begin natural gas development from their jointly owned offshore *Dorra Gas Field*, noted Kuwait's Minister of Energy, Sheikh Ahmed Fahed al-Sabah, on 2 October 2003. The decision to develop the controversial *Dorra Gas Field*, in the Partitioned Neutral Zone (PNZ), has implications for relations with Iran, which also claims part of the field and whose maritime border with Kuwait remains undefined. Sheikh Ahmed made no mention of whether Kuwait had resolved the maritime boundary delimitation issue with Iran, which was to have been included in the sharing proposal of the reserves. Kuwait and Saudi Arabia delimited their sea boundary with an agreement in July 2000, which resulted in the equal sharing of Dorra Field within the PNZ. Kuwait and Iran have been discussing their offshore boundary since 2000, the same year Iran halted a drilling programme it had quietly started in the crest of the Dorra field, at the request of Kuwait and Saudi Arabia. Kuwait had earlier indicated that it would wait for the delimitation talks with Iran to be concluded before starting any developmental work at the field. Kuwait has also stated that it would offer a portion of the field to Iran in the agreement. Iranian officials are of the opinion that the remarks by the Kuwaiti Minister were designed to put pressure on Iranian authorities to speed up agreement on boundary delineation. The two countries have signed preliminary agreements for Iranian gas sales through a proposed pipeline to Kuwait. The Kuwaiti Deputy Energy Minister, Isa al-Oun, observed that this project remained a priority.

Fishing: an International and Regional Problem

While India and Pakistan bicker over the maritime boundary, the local fishers who have traditionally operated the north-eastern waters of the Arabian Sea are apprehended if they stray over the 'disputed line'. Neither the plentiful fish nor the fishers in the area respect jurisdictional limits, especially if these lines are not delineated. The Indian Coast Guard and the Pakistani Maritime Security Agency are out to prove their respective points. Both agencies routinely apprehend fishing vessels from the other side and punish the fishers, sometimes for years. Every so often, when possibly the jails are full or when the Authorities deem necessary, the fishermen are released with great publicity. Prior to the Agra Summit, the Indian government released all Pakistani fishers and gave orders to the Coast Guard to resort only to warnings instead of apprehending the erring crew. There has been no reciprocity from the Pakistani side and the problem continues to fester.

In its effort to protect the Patagonian Toothfish (*Dissostichus eleginoides*) from illegal fishing in the southern waters of the Indian Ocean, Australia on 23 October 2002, sought to have the deep sea species protected under international law. Earlier in the year, Australian Environment Minister, Dr. David Kemp, announced that his country would nominate the Patagonian Toothfish under Appendix II of the UN Convention on International Trade in Endangered Species (CITES), when Parties to the treaty meet in Chile, on 3-15 November 2003. A CITES Appendix II listing applies to species that are not necessarily threatened with extinction, but that may become so unless trade is controlled. Species listed under Appendix II must be accompanied by an appropriate CITES export permit issued by the exporting country before entry to the importing country will be allowed. Appendix I, on the other hand, includes species threatened with extinction. According to Australian government officials, a total of 200 metric tons of Patagonian Toothfish, worth about US$1.3 million, was confiscated from the two ships. The ships and crew members were held in custody pending a court hearing in November 2003. In the meantime, the Supreme Court of Western Australia has recently increased the bail for the Russian captain and three Spanish crew of the *Volga* from US$37,500 to US$137,500 per person, while the three crew members from the *Lena* were fined US$50,000. Those in custody were not permitted to leave Australian jurisdiction until cash deposits for the bail were received. The Australian Fisheries Minister noted that the initial apprehension of the two Russian flagged vessels was proof of his Government's determination to protect Australia's fishing interests and sovereignty and to preserve the stock of valuable species, such as the Patagonian toothfish.

As a result, the Australian government has increased surveillance and enforcement of the waters surrounding Heard and McDonald Islands to minimize illegal catches of the valuable toothfish. The fisheries around Heard and McDonald Islands, where the two ships were caught, were worth some US$15 million a year to legal Australian fishing operators, according to the Australian Fisheries Management Authority. However, its remoteness creates logistical problems for Australian protection, which illegal fishing boats often try to exploit. Other territories in the region, such as the French island of Kergeulen, have faced similar incursions by illegal operators. This is not the first time foreign trawlers have been caught fishing illegally in Australian waters.

A Togo registered vessel, the *South Tomi*, was caught with a US$800,000 haul taken from the Australian Fishing Zone around the Australian territory of Heard Island in the southern Indian Ocean. The *South Tomi* was captured in March 2001, after a 14 day hot pursuit across the Southern Ocean involving three countries – France, South Africa and Australia. Its captain has been ordered to pay US$68,000, the highest fine imposed in Australia for poaching to date. It appears that the seizure of the two Russian-flagged vessels was just the tip of the iceberg. According to Greenpeace Australia, the *Lena* and the *Volga* are but two of many pirate fishing vessels poaching the Patagonian toothfish. A report by the wildlife trade monitoring organization, *TRAFFIC*, stated that the total trade in unprocessed toothfish for 1999-2000 was as high as 59,000 metric tons. Up to 33,000 tons of this amount were caught by pirate ships like the *Lena* and the *Volga* and sold in poorly regulated ports in such countries as Mauritius, Namibia, Uruguay, and, increasingly in Indonesia.

Like other deep-sea fish, very little is known about the Toothfish. It is known that they have a lifespan of up to 50 years, but do not start spawning until they are 10 to 12 years of age. Toothfish are one of the two largest species of fish occurring in the Antarctic, reaching up to 2.2 meters (7 feet) in length and up to 100 kilograms (220 pounds) in weight. As a result of illegal fishing, particularly the overfishing of juvenile fish, many scientists believe that the Patagonian Toothfish will be commercially extinct in several years, very much like the marbled rockcod and the mackerel icefish. After just two years of pirate fishing of Patagonian toothfish around Crozet Island, southeast of South Africa, the fishery has already reached the point of commercial extinction.

Marine Terrorism and Acts of Piracy

The vulnerability of shipping to terrorist attacks is highlighted in a report on piracy and other criminal attacks at sea issued by the ICC International Maritime Bureau (IMB). The IMB Director, Captain Mukundan, noted that "Patrols should be intensified and naval and police units should immediately board and investigate any unauthorized vessels caught loitering in the no-go areas". The responsibility for security of waterways must remain with the governments concerned. The appeal came amid mounting evidence that the blast that ripped through a French tanker, the *Limburg*, earlier in early-October 2003, in the Gulf of Aden, was a terrorist attack. The Limburg is thought to have been rammed by a small boat noticed by crew members shortly before the explosion. Captain Mukundan noted that no shipboard response could counter that kind of attack.

The IMB advised ships calling at Middle Eastern, Pakistani and Indonesian ports to take advice from local shipping agents and owners about the degree of risk involved. The identity of the attackers of the *Limburg* was allegedly from Yemen. The tanker was carrying 400,000 barrels of crude oil when the blast ignited a fire on board, killing one crew member and spilling 90,000 barrels of oil into the sea. Immediately, the French Defence Minister, Michele Alliot-Marie, announced that parts of a small boat and traces of TNT were found inside the tanker.

The IMB *Annual Piracy Report* for 2002 stated that attacks like the one in the Gulf of Aden in October 2002, were difficult to prevent, noting that no shipboard response can protect the ship in these circumstances. The only answer was for coastal states to make sure that approaches to their ports were secure. IMB recommended that port authorities designate approach channels under coast guard or police supervision from which all unauthorized craft would be banned. "The risk of terrorist attack can perhaps never be eliminated, but sensible steps can be taken to reduce the risk" the IMB said. At issue here is how seriously do the governments, in the region and world-wide take the threat of maritime terrorism.

Acts of Piracy and Armed Robbery, however defined, were 370 during 2002 – an increase from 335 in 2001. Indonesia again experienced the highest number of attacks, with 103 reported incidents in 2002. In the first-half of 2003, there were at least 150 incidents in the Indian Ocean region, of which 64 was recorded in Indonesian waters. Overall, most attempts and attacks occurred while ships were at anchor. A marked increase in successful boarding by pirates combined with a drop in the number of attempted attacks suggested that many ships were complacent about the need for additional precautionary measures. There was a substantial rise in hijackings, up from 16 to 25 incidents. Many involved smaller boats, such as tugs, barges and fishing boats, in the Malacca

Straits and Indonesian waters. Crime syndicates in the area were believed to be targeting vessels carrying valuable palm oil and gas oil in the Southeast Asian region. A general consensus suggests that, in some parts of the world, it is all too easy to board a merchant vessel unlawfully. Gangs of heavily armed pirates using fishing and speed boats have been targeting small oil tankers in the Malacca Straits, according to the ICC's Piracy Reporting Centre. Issuing fresh warnings weekly to commercial ships operating in the area, the IMB warned that the recent wave of attacks follow a pattern set by Indonesian Aceh rebels. Captain Pottengal Mukundan of the ICC's International Maritime Bureau (IMB) – whose organization manages the Piracy Reporting Centre – says that the latest attacks raise a number of serious concerns. "In addition to the obvious threat to human life and potential environmental damage, we are very concerned about politically motivated attacks against vessels". Captain Mukundan observed that there was evidence to suggest that Aceh rebels are responsible for the growing piracy in the area. Their principal motivation, he said, is to fund their political cause by holding hostages for ransom. Political piracy threatens to rewrite the rules of engagement and authorities need to recognize the motives behind these crimes and adopt new methods of tracking and deterring them.

Although the number of crew killed in 2002 was down to ten compared with 21 in 2001, that figure concealed a chilling statistic – 24 passengers or crew were missing, and most of these must be considered dead. The report's summary of attacks on ships frequently noted that pirates threw crew members into the sea, leaving them to drown. Piracy attacks in Bangladesh ranked second highest with 32 attacks, and India was third with 18 attacks.

The waters off Somalia are among the most dangerous in the world. The risk of attack to vessels staying close to the coastline from Somali armed militias had increased from one of possibility to certainty. Any vessel not making a scheduled call in a Somali port which slows down or stops close to the Somali coast could well be boarded by these gangs. The gangs have extorted substantial sums from owners of ships for the return of the vessel and crew. Other problematical geographical areas include the northern and southern waters of the Malacca Straits, and the southern approaches to the Red Sea.

The IMB Piracy Reporting Centre in Kuala Lumpur, which runs a satellite warning system for ships at sea, provides assistance free of charge to ships that have been attacked. A weekly summary of the Centre's daily satellite reports is posted on the Internet at www.icc-ccs.org.

Conclusion

Some states in the Indian Ocean region have demonstrated their desire to complete maritime boundary negotiations with their neighbours in order to progress. Given the will and the determination by the other governments in the region to put aside terrestrial geopolitical issues and negotiate on the concerns and issues that relate to their adjacent seas, agreements can easily be reached to delimit their respective maritime boundaries – the remaining fifty per cent or so, to complete the political division of the Ocean. Similar agreements can be forged to assist each other in search and rescue at sea, to prevent pollution, and for joint action against acts of piracy and armed robbery and marine terrorism. The scope is unlimited. In so doing, they could avoid costly visits to international tribunals, expedite resource development projects and importantly create joint development zones, as an interim measure, if disputes cannot be resolved easily.

A more pragmatic and humane method must also be initiated in dealing with fishers who are apprehended by authorities who themselves are unsure of national jurisdictional limits in the region. Other jurisdictional issues include the laying of oil and gas pipelines, avoiding incidents at sea between naval ships of different countries, and cooperation in search and rescue in the marine environment. Vigilant anti-piracy watch is still the best deterrent when it concerns acts of piracy and armed robbery. To overcome current concerns in respect of maritime terrorism, it is vital that coastal states allocate resources to patrol their waters more effectively. Failure to do this may well result in an increased number of incidents.

References

Attard, D.J. (1987), *The Exclusive Economic Zone in International Law* (Oxford: Clarendon Press).

Bernaerts Arnd (1988), *Bernaerts' Guide to the Law of the Sea* (Coulsdon, Surrey: Fairplay Publications).

Charney J. I. and Alexander L. M., eds. (1993), *International Maritime Boundaries*, Volumes I, II and III, (The Hague: Martinus Nijhoff Publishers).

Charney, J. I.. and Smith, R. W., eds. (2002), *International Maritime Boundaries*, Volume IV, (The Hague: Martinus Nijhoff Publishers).

Forbes, V. L. (1992), "British Indian Ocean Territory: Chagos Archipelago", *Indian Ocean Review*, 5, (1), 16-19.

Forbes, V. L. (1995), *The Maritime Boundaries of the Indian Ocean Region* (Singapore: Singapore University Press).

Forbes, V.L. (2001) *Conflict and Cooperation in Managing Maritime Space in Semi-enclosed Seas*, Singapore: Singapore University Press.

Kwaitkowska, Barbara (1989), *The 200-Mile Exclusive Economic Zone in the New Law of the Sea* (The Hague: Martinus Nijhoff).

Prescott, J. R. V. (1985), *The Maritime Political Boundaries of the World* (London: Methuen).

Smith, R.W. (1986) *Exclusive Economic Zone Claims*, Dordecht: Martinus Nijhoff

United States – Department of State – Office of the Geographer *Limits in the Sea Series*.

United Nations Division for Ocean Affairs and Law of the Sea
http://www.un.org/Depts/los/index.htm

International Chamber of Commerce – International Maritime Bureau Piracy Reporting Centre

http://www.iccwbo.org/ccs/news_archives/2003/piracy_ms.asp

APPENDIX 11.1

Status of The United Nations Convention on the Law of the Sea, 1982
The 1982 Convention

Ratifications of, accessions and successions to the Convention and the related Agreements, as at 13 August 2003.

1ˢᵗ State   Fiji (10 December 1982)
143ʳᵈ State (latest)   Albania (23 June 2003)

Agreement Relating to the Implementation of Part XI of the Convention
1ˢᵗ State   Kenya (29 July 1994)
115ᵗʰ State (latest)   Honduras (28 July 2003)

Agreement for the Implementation of the Provisions of the Convention relating to the Conservation and Management of Straddling Fish Stocks and Highly Migratory Fish Stocks

1ˢᵗ State   Tonga (31 July 1996)
36ᵗʰ State (latest)   India (19 August 2003)

APPENDIX 11.2

INDIAN OCEAN REGION
MARITIME JURISDICTIONAL LIMITS CLAIMED BY STATES

Column 1 Common short-form name of country
Column 2 Territorial Sea limit claimed by the State
Column 3 Most recent date of Baseline declaration
Column 4 Limit of Contiguous Zone claimed by the State
Column 5 Limit of Exclusive Economic Zone
Column 6 Limit of Continental Shelf natural or legal
Column 7 Number of maritime boundaries delimited
ND = not defined   NM = nautical miles

Common Name	TS NM Determined	Baseline Declared	CZ NM	EEZ NM	CS metres	MB
Australia 12	4-2-83	24	200	200/CM	8	
Bahrain	12	–	24	–	–	3
Bangladesh	12	13-4-74	18	200	CM	ND
Comoros	12	6-5-82	–	200	–	ND
Djibouti	12	5-5-85	24	200	–	ND
East Timor	12	–	–	200	–	1
Egypt	12	9-1-90	24	200	CS	ND
Eritrea	12	15-9-91	–	–	–	1
France	12	19-10-67	24	200	CS	16
India	12	–	24	200	200NM	5
Indonesia	12	18-2-60	–	200	defined	13
Iran	12	21-7-73	24	200	CS	5
Iraq	12	–	–	–	–	1
Israel	12	–	–	–	CS	1
Jordan	3	–	–	3+	–	1
Kenya	12	16-5-72	–	200	CS	1
Kuwait	12	24-12-67 –	–	–	1	
Madagascar	12	27-2-63	24	200	200	ND
Malaysia	12	–	–	200	defined	6
Maldives	12	27-6-96	24	200	defined	2
Mauritius	12	16-4-70	–	200	200nm/CM	1
Mozambique	12	18-8-76	–	200	–	1
Myanmar	12	9-4-77	24	200	200NM	2
Oman	12	1-6-82	24	200	CS	1
Pakistan	12	29-8-96	24	200	200NM	ND
Qatar	12	–	24	–	CS	2
Saudi Arabia	12	16-2-58	18	–	defined	3
Seychelles	12	12-3-99	24	200	200	ND
Singapore	3	–	–	3+	–	2
Somalia	200	10-9-72	–	–		ND
South Africa	12	11-11-94	24	200	CM	ND
Sri Lanka	12	15-1-77	24	200	200NM	2
Sudan	12	31-12-70 18	–	CS	2	
Tanzania	12	15-10-89 –	200	–	2	

Common Name	TS NM Determined	Baseline Declared	CZ NM	EEZ NM	CS metres	MB
Thailand	12	17-8-92	–	200	CS	4
United Arab Emirates	12	17-10-93	24	200	CS	8
UK (BIOT)	3	15-11-98	–	200	CS	ND
Yemen	12	15-1-78	24	200	200NM	1

South Africa Claims a 24-nautical mile 'maritime cultural zone'; claims same rights and powers over archaeological or historic items as in the territorial sea.

Many of the States stipulate that foreign warships must obtain permission prior to entering their territorial sea

Sources: UK Hydrographic Office (HH. 085/012/01) and author's research

APPENDIX 11.3

MARITIME BOUNDARY AGREEMENTS BETWEEN INDIAN OCEAN LITTORAL STATES

The following maritime boundary agreements, or adjudications, are in force unless followed by an asterisk (*) indicating that the agreement has been signed only. Unless otherwise indicated the delimitation creates a maritime boundary separating all applicable maritime jurisdiction permitted under international law, including the territorial sea, fishing, continental shelf, and other rights permitted in the exclusive economic zone. Boundaries separating other specific types of jurisdiction are noted as follows:

TS = territorial sea; CZ = contiguous zone; CS = continental shelf; F = fishing zone.

The date that the agreement or adjudication entered into force, or was signed, is in parenthesis.

> Australia (Heard/McDonald Islands)/ France (Fr. Southern and Antarctica Lands - (9 Jan 1983)
> Australia/East Timor (21 May 2000)
> Australia/Indonesia (2 boundaries; CS; 8 Nov 1973)
> Australia/Indonesia (F; 1 Feb 1982)
> Australia/Indonesia * (CS; 11 Dec. 1989)
> Australia/Papua New Guinea (2 boundaries; 15 Feb 1985)
> Bahrain/Iran (CS; 14 May 1972)
> Bahrain/Saudi Arabia (CS; 26 Feb 1958)
> Bahrain/Qatar (ICJ Decision 16 March 2001)
> Burma/India (CS; 14 Sept 1987)
> Burma/Thailand (CS; 12 Apr 1982)

Dubai/Sahrjaj (Arbitral Award 19 October 1981)
Eritrea/Yemen (Arbitral Award 17 December 1999)
France (Reunion)/Mauritius (CS; 2 Apr 1980)
France/Seychelles
India/Indonesia (CS; 17 Dec 1974 and 15 Aug 1977, two agreements)
India/Indonesia/Thailand (CS; 2 Mar 1979 – tri-point agreement)
India/Maldives/Sri Lanka (CS; 31 July 1976 – tri-point agreement)
India/ Maldives (CS; 8June 1978)
India/Sri Lanka (CS; 8 July 1974 and 10 May 1976, two agreements)
India/Thailand (CS; 15 Dec 1978)
India/Thailand (27 October 1993)
India/Myanmar /Thailand (24 May 1995)
Indonesia/ Malaysia (CS; 7 Nov 1969)
Indonesia/Malaysia (TS; 8 Oct 1971)
Indonesia/Malaysia (2 boundaries; CS; 7 Nov 1969 and 16 July 1973, two agreements)
Indonesia/Malaysia/Thailand (CS; 16 July 1973 – tri-point agreement)
. Indonesia/Papua New Guinea (2 boundaries; TS; 26 Dec 1974)
Indonesia/Papua New Guinea (CS; 10 July 1982)
Indonesia/ Singapore (TS; 29 Aug 1974)
Indonesia/Thailand (CS; 16 July 1973 and 15 Dec 1978, two agreements)
Iran/Oman (CS; 28 May 1975)
Iran/Qatar (CS; 10 May 1970)
Iran/Saudi Arabia (CS; 29 Jan 1969)
Iran/United Arab Emirates (Dubai) * (CS; 31 Aug 1974)
Iraq/Kuwait (21 May 1993) [UN Iraq/Kuwait Boundary Demarcation Committee]
Kenya/Tanzania (9 July 1976)
Kuwait/Saudi Arabia (30 January 2001)
Malaysia/Singapore (TS; 3 Aug 1928)
Malaysia/Thailand (TS; 15 July 1982)
Malaysia/Thailand (CS; 16 July 1973 and 15 July 1982, two agreements)
Mozambique/Tanzania * (28 Dec 1988)
Oman/Pakistan (signed 11 June 2000)
Qatar/United Arab Emirates (Abu Dhabi - CS; 20 Mar 1969)
Saudi Arabia/Yemen (9 July 2000)
Saudi Arabia/Sudan (Common Zone)
Thailand/Vietnam (27 December 1997)

CHAPTER 12

Contemporary Piracy, Terrorism and Disorder at Sea: Challenges for Sea-Lane Security in the Indian Ocean

Vijay Sakhuja

In the wake of the tragic events of 11 September 2001, in the United States, the International Maritime Organization (IMO) called for a review of measures to combat violence and crime at sea. (IMO's Home page). A resolution calling for the review of existing international legal and technical measures and procedures to prevent acts of terrorism that threaten the security of ships at sea and ports was also adopted. It was agreed to adopt new regulations and mandatory statutory instruments to enhance maritime security by preventing and suppressing acts of terrorism against shipping and make sea-lanes safe for international commerce. Meanwhile, the UNCTAD report 'Review of Maritime Transport 2001' had noted that, world-wide, the merchant fleet continued to expand at 1.2% and, at the beginning of 2001, was pegged at 808.4 million dead weight tonnage (dwt). (UNCTAD Home page) During 2000, 44.4 million dwt (up 9.6% from 1999) of new ships were delivered, 22.2-million dwt (down 27.7%) tonnage were broken up and 12.8 million dwt were removed from inventories due to vessel losses/decommissioning, resulting in a net gain of 9.4 million dwt. Furthermore, world sea-based trade recorded its fourteenth consecutive annual increase, and Asia's share of imports and exports were 26.1 and 18.8 percent, respectively. Given these trends, sea-based commerce has entered a mood of great optimism.

While this optimism is a welcome sign, sea-lanes continue to be vulnerable to disruption. Piracy, terrorism, drug trafficking, gun running, human smuggling and pollution are threatening these routes. None of these activities are independent of the other and failure in one often leads to failure in others. In this regard, it is argued in this Chapter that the sea-lanes in the Indian Ocean are threatened from several directions and that there is an urgent need for greater cooperation among littoral states to enhance security and safety of the maritime enterprise.

Sea–Lanes in the Indian Ocean

The Indian Ocean is home to important sea-lanes. A large volume of international long-haul maritime cargo from the Persian Gulf, Africa, the Asia-Pacific and Europe transits through its waters. It is home to important Straits: Hormuz, Malacca, Sunda, Lombok and Suez. Hormuz lies between Oman and Iran. It is strategically important to most countries of the world. The primary cargo through the Strait is oil/gas, which virtually affects every aspect of the daily lives of most people of the world. It is fairly deep and vessels of 1,60,000 dead weight tonnage can pass through the waterway and nearly 15.5 million barrels of oil flows through it daily (Tatsuo Masuda: Planetark Webpage, 2002). From the Persian Gulf, the sea-lane transits through the Indian Ocean almost hugging the Indian coast towards the Strait of Malacca. Malacca is the busiest with over 600 vessels transiting through it every day. Each day, about 10.3 million barrels are carried through the Strait of Malacca. Virtually all ships destined for East Asia carrying LNG and LPG pass through Malacca and the issue of safety is likely to grow in importance as East and South East Asia's energy imports grow. It is estimated that the number of tankers transiting through Malacca would increase to 59 per day in 2010 from 45 in 2000. Similarly, the LPG tanker traffic is expected to increase to seven per day in 2010 from five in 2000 and LNG tankers are expected to rise to 12 per day from eight in 2000.

The sea-lane also runs towards the west to pass through the Suez Canal into the Mediterranean Sea. The Suez Canal transported around 820,000 bbl/d of petroleum in 2000, and, in 2001, about 360 oil tankers passed through the canal. Currently, VLCCs and ULCCs cannot pass through it but the Egyptian government plans to widen and deepen it by 2010 to allow these large vessels to transit. The SUMED pipeline, linking the Ain Sukhna terminal on the Gulf of Suez with Sidi Kerir on the Mediterranean has a transport capacity of 3.1 million bbl/d. Closure of the Suez Canal and/or the SUMED Pipeline would divert tankers around the Cape of Good Hope, adding to transit time and tanker capacity (Webpage of EIA). It is abundantly clear that the Straits of Hormuz, Malacca and Suez are strategically important to the Indian Ocean sea-lane security.

Threats to Sea-lanes in the Indian Ocean

Threats to sea-lanes arise from several directions. Some of these can be attributed to external factors and some to internal, while there are other factors (storms, cyclones, typhoons, fog, rain) over which the maritime community has limited control. It is therefore important that threats to sea-lanes be classified into two distinct groups: external and internal. This classification of threats helps the maritime community to develop sophisticated strategies to ensure the safety and security of sea-lanes. External factors would include piracy and terrorism, while the internal factors are partly the making of the maritime community itself – that is, Flag of Convenience (FOC) shipping, incompetent crew, poor seamanship and illegal activities like gun running, drug smuggling and illegal transportation of humans by ships.

External Factors

Piracy – Piracy at sea continues to top the list of external threats to sea-lanes. According to the International Chamber of Commerce Annual Report on Piracy at Sea, 2001, there were 335 attacks on ships in 2001. Pirates killed 21 crew members/passengers, and 210 seafarers were taken hostage during the attacks (IMB Webpage, February 2002). A new trend of 'kidnap and ransom' piracy was also observed in the Malacca Straits (waters around Aceh, Indonesia), a phenomenon earlier restricted to Somali waters. During 2000, piracy attacks had shown an alarming rise, with armed robbery in the waters around Indonesia, Bangladesh, the Malacca Straits, India and the Red Sea. Similar trends continued during 2001, with Indonesia, India and Bangladesh occupying the top three positions with 91, 27 and 25 attacks, respectively.

The Malacca Strait and the Singapore Strait appear to be popular among pirates. This is due to geographical and operational reasons. The area around the Straits attracts the heaviest maritime traffic concentration. As noted earlier, over 600 vessels of different types transit the Malacca Strait every day, and the traffic density is very high. Although there is a traffic separation scheme, transiting vessels reduce speed to negotiate the traffic. Very Large Crude Carriers (VLCCs) transit the strait at safe speeds with only a meter or two clearance under the keel. In addition, the Strait is home to several shipwrecks and shallow areas. These navigational and operational conditions provide the right type of environment for pirates to undertake attacks as well as facilitating a quick getaway. The number of attacks is fast increasing and is alarming.

In the recent past, the waters off Sri Lanka have witnessed frequent acts of piracy. In one incident, a merchant vessel, M V Cordiality, was captured and five Chinese crew were allegedly killed by Sri Lankan terrorists near the port of Trincomalee (Xinhua, 11 September 1997). LTTE rebel forces are also reported to have hijacked ships and boats of all sizes, and kidnapping and killing crew members is a common practice. In August 1998, a Belize flagged general cargo vessel, M V Princess Kash, was hijacked by LTTE rebels (Gunaratna, 1998). While on its way to Mullaitivu, a LTTE stronghold, the Sri Lankan Air Force bombed the vessel to prevent the ship's cargo falling into the hands of the LTTE. The status of the 22 crew members is still not known. The International Maritime Bureau (IMB) has warned commercial shipping transiting along the east coast of Africa not to come within 50 nautical miles of the coastline because of pirates. Most of the ports in the Horn of Africa are piracy-prone and are dangerous. Attacks occur in port, at anchor or during slow-steaming off the coast. Somali waters, in particular, have been notorious for acts of piracy. This is primarily due to political instability. Somali militias operate speedboats and are reported to pose as coastguards. Their activities include kidnappings, vessel seizures and ransom demands. The most dangerous aspect of these activities is the frequent use of infantry weapons like mortars, grenades and small arms.

In January 2002, a Lebanese vessel, *Princess Sarah*, was attacked and hijacked twice off the coast of Somalia (Piracy Quarterly Reports: Webpage wanadoo). In the first instance, the vessel was fired at while underway. In the process, it sustained some damage and was forced to stop after 20 miles to make repairs to the engine. Two days later, a second attack took place and the pirates managed to take over the vessel. The initial ransom demand was for $60,000, but this was then changed to $200,000. An agreement was reached on an undisclosed amount and the crew was released. A French navy vessel was sent to oversee the release. The troubles of the vessel the did not end there because another attack took place and this time the pirates were scared away by a helicopter that was sent by the nearby French warship, the Floreal. In June 2002, the Cyprus-flagged, Panagia Tinou, a bulk carrier, was hijacked while it was at anchor off the North coast of Somalia. The crew was taken hostage and the ship was moved to an unknown destination along the Somali coast. The United Nations entered into negotiations with the pirates on behalf of the Philippines, since Somalia has no bilateral relations. After 16 days of captivity, the crewmen were released in return for a ransom of $400,000, paid by an international ship insurance agency through the Protection and Indemnity Club. According to the commander of the German Naval Air Wing Detachment in Mombasa, the presence of three German vessels near where the hijacked vessel was anchored helped free it.

Pirates are like sharks. They breed and show up more often in some regions than others. They can strike anywhere, but some areas are popular with them. Unfortunately, the Indian Ocean is home to the top three piracy-infested regions. South East Asia, South Asia and the East African coast are hotspots of piracy and account for more than half of the world-wide reported attacks. It appears that piracy moves around and its centre of gravity keeps shifting, but pirate-plagued areas remain active. It may simply happen that one area may record more attacks than the others, but no area can claim itself to be free from pirates. The number of incidents including violent attacks continues to rise. The maritime community is growing pessimistic about the ability of states to counter piracy. They also appear to doubt the intention of states to take the matter seriously.

Terrorism – The threat of terrorism to sea-lane security is not a new phenomenon. In the past there have been several instances when ships have been taken over by insurgents or attacked by terrorists. (Menefee, 1991, 43-55). The takeover of a Portuguese passenger liner, Santa Maria, in 1961, ushered in the modern age of maritime terrorism. In 1988, the City of Poros, a Greek cruise ferry carrying 500 tourists, was attacked by terrorists. However, it was the Achille Lauro incident that caught the attention of the international community with regard to the menace of terrorism at sea and the formulation of an international convention under the United Nations. Similarly, on November 03, 1988, two trawlers carrying 150 PLOTE (People's Liberation Tamil Tigers Eelam) mercenaries landed in the Maldives (WebPages of Bharat-Rakshak). An Indian Navy maritime reconnaissance aircraft detected the ship and Indian navy vessels later captured it.

In recent times, the USS Cole, an Arleigh Burke-class destroyer, was attacked by terrorists: In October 2000, during its refueling halt in Aden , Yemen, a small craft packed with explosives crashed into the warship and exploded, leaving it with a forty foot hole on its side and several crew members dead. The bombers had links with Al-Qaida and America's Number 1 terror suspect, Osama bin Laden (Webpage of Cargolaw). As part of the campaign against terrorism, the Singapore authorities arrested 15 suspected Islamic militants with links to Al-Qaeda who were planning to blow up U.S. naval vessels and a bus that was to transport American military personnel (Webpage of Pilotonline). These incidents have also exposed the vulnerability of warships.

The recent terrorist attack off Yemen on the French oil tanker, Limburg, is indeed a chilling reminder of the vulnerability of maritime enterprise to asymmetric threats. The owners of the French-flagged tanker believe that a speedboat packed with explosives rammed into the ship, leaving it disabled and leaking oil. This suits the terrorist requirements of low technology warfare against a superior force. It is thus clear that the threat of terrorism to commercial maritime enterprise is far from rare and is increasingly becoming a reality. Unfortunately, maritime shipping is the soft underbelly of states that offers the best form of vehicle to carry out illegal activities.

Shipping Containers – It is now increasingly accepted that shipping containers are an efficient delivery system for nuclear, chemical and biological weapons of mass destruction. Large container vessels dock at ports almost everyday. These vessels are subjected to routine crew and container manifest inspections. The container itself is rarely subjected to a thorough inspection. Although the introduction of shipping containers in 1956 was a revolution in the process of shipping vast amounts of goods across the ocean, these boxes, which can be easily shifted to railroad cars or trucks, have emerged as an important tool for terrorists. Modern container vessels like the Regina Maersk can transport as many as 6600 TEU containers.

After the container door is closed and sealed (under varying degrees and seriousness of inspection), these boxes move into seaport terminals, aboard container ships, trains, trucks and even aircraft, with little or no information about their true contents. As regards containers, only a few are thoroughly examined for their content. According to maritime security experts, containers are the most suitable means of transporting men and materials for terrorist activities including weapons of mass destruction.

Meanwhile, the freighter Palermo Senator, was ordered to remain six miles off the coast of New Jersey. This vessel was scheduled to offload 655 containers in Port Elizabeth. The US Coast Guard agents detected traces of low radiation (Web Page of DEBKA – Net-Weekly Intelligence Report, June 2002) while the ship was being checked for stowaways. The vessel was registered under the Liberian flag and owned by a German company. Its last port of call was Valencia, Spain, although the ship had previously called at various ports in Southeast Asia and transited through the Suez Canal.

Earlier, a suspected Al-Qaeda operative smuggled himself halfway around the world locked inside a shipping container with his own bed and toilet. Apparently he was carrying security passes and maps of some airports in Canada, Thailand and Egypt. A laptop computer, two mobile phones, cameras, a Canadian passport, other identity documents and a certificate saying he was an aircraft mechanic were found on his person. According to US customs authorities, only 2 per cent of the cargo containers that enter seaports each day are inspected. Similarly, in New Orleans, a container labeled as empty held oil exploration tools that were radioactive. When customs officials opened the container in the port, the beeper-sized radiation alarms on their belts screamed a warning. The inspectors had to summon a decontamination team (New York Times , December 2001).

As part of maritime security initiatives, the United States has sought cooperation from several countries to improve security at ten "mega-ports" that account for nearly half of all containers that are landed at US seaports. Among other issues, this included placing US custom inspectors to identify high-risk containers and screening them at the port of origin. According to bilateral agreements with Canada, US customs inspectors have now been stationed at three Canadian seaports and also in Singapore and Hong Kong.

Mining of Choke Points – At present, there are no indications to suggest that choke points are likely to be threatened by states astride these waterways, but such eventualities cannot be rued out. Similarly, the threat of mining by regional terrorist organizations to draw international attention remains a potent threat. There are distinct possibilities of disruption of sea-lanes due to interstate conflicts. States are bound to declare maritime exclusion zones or areas dangerous to shipping due to conflict at sea. Although countries have been pursuing an aggressive maritime cooperation strategy both bilaterally and on a multilateral basis to prevent scuttling of sea-lanes, conflicts have continued to threaten shipping.

Internal Factors

'Flag of Convenience' Shipping – The world's 'most wanted terrorist', Osama bin Laden and his Al Qaeda operatives, are known to own or have chartered at least 20 merchant vessels capable of undertaking ocean passage. These vessels are suspected to possess a 'flag of convenience' (FOC) registry in Liberia, Panama and the Isle of Man (Webpage of Amo-Union, October 2001). It is believed that Osama's secret shipping fleet, flying a variety of flags of convenience, allows him to hide the ownership of vessels, transport goods, arms, drugs and recruits with little official scrutiny. A shipbroker in Germany admitted acting as a translator when Wahid al Hage, an Al Qaeda operative, sought to buy a merchant vessel. Wahid is sought in connection with the 1998 bombings of two U.S. embassies in East Africa. Reportedly, one of bin Laden's cargo freighters unloaded supplies in Kenya for the suicide bombers who weeks later destroyed the U.S. embassies in Kenya and Tanzania.

The presence of FOC (Flag of Convenience) vessels has indeed exposed chinks in the armour of security and poses new challenges to maritime forces. Several terrorist organizations are known to possess merchant ships. For instance, the LTTE has a flotilla of ships that are engaged in maritime trade. Most of these are registered in FOC countries known as " pan-ho-lib" – that is, Panama, Honduras and Liberia (*Sunday Times*, Colombo, February 1996).

Operating under flags of convenience and staffed with a Tamil crew, these cargo ships carry out their activities between Asia and Europe. They are also known to regularly transport illegal immigrants or to be involved in any kind of traffic on behalf of the LTTE. For instance, LTTE vessels ship narcotics from Burma to Turkey. In order to give legitimacy to their fleet and also to generate revenue, LTTE ships move general cargo, such as fertilizers, timber, rice, cement and other dry cargo. (Gunaratna, 1999, 266).

The capture by the Israeli Navy of Karine-A, a Palestinian Authority ship, carrying 50 tons of lethal ammunition, offers a classical example of a change of name. According to Lloyd's of London, the vessel was owned by the Beirut based Diana K Shipping Company and was registered in Lebanon. Its original name was Rim K and was purchased by Ali Mohammed Abbas, an Iraqi national, for $400,000. It was re-registered in Tonga as the Karine-A (WebPages of Stratfor, January 2002).

According to the International Transport Worker's Federation (ITF) Fair Practices Committee, a union of seafarers and dockers campaigning against FOC, there are thirty countries that are known to offer FOC registry (WebPages of ITF). These are essentially developing or small island states. The ITF believes that there should be a genuine link between the vessel and its flag, which would increase accountability and force ship owners to maintain international shipping standards, a practice that does not commonly occur with FOC ships. According to industry experts, flag hopping is a common practice and ship owners tend to switch registry at the first sign of a crackdown by authorities or when engaging in activities involving gun running, drug smuggling, transporting illegal cargo or human beings.

FOC vessels have the capability to challenge sea-lane security. Their presence clearly highlights the dangers involved when strategic cargo like oil is being transported. It is also a reminder that countries need to revitalize national fleets and challenge FOC vessels. If the maritime community is serious about challenging the forces of terrorism, then it must establish a framework for a genuine link between the flag a vessel flies and the state to which it belongs.

Crew Competence – Though merchant ship crews are trained to handle emergencies on board ships, they have been found wanting in professional competence. According to the International Maritime Bureau (IMB), it is virtually impossible to verify the authenticity of the identity of the crew. Besides, there is a major problem of counterfeit and improperly issued mariner documentation. The IMB has issued a warning to ship operators about the thousands of unqualified crew and masters working illegally with false papers, and has called for tighter security by authorities issuing certificates.

The alert follows the release of statistics showing that, of 54 maritime administrations surveyed, more than 12,000 cases of forged certificates of competency were reported. These figures highlight the gravity of the situation. Ships are sailed by crewmen with false passports and competency certificates. The IMB also believes that, at times, the issuing authorities themselves are to blame. For instance, the Coast Guard office in Puerto Rico was reported to have issued nearly 500 suspicious certificates of competency. Such cases usually escape detection by the port authorities.

The ships' crews themselves can act as a potential threat. Both the Philippines and Indonesia are the largest suppliers of merchant ship crews. These states are home to radical groups like Abu Sayyaf and the Free Aceh Movement (GAM). Under the circumstances, it is virtually impossible to detect potentially undesirable crew members. The situation becomes more complicated in the case of vessels which fly 'flags of convenience' and employ multinational crews. It is difficult to verify the authenticity of the identity of the crew.

Interestingly, the ship itself is a safe haven for undesirable elements. There are several hidden spaces, holds and compartments in the ship that are difficult to inspect. Some spaces are so unfamiliar that it may be difficult to locate them without the help of the ship's compartment drawings. Modern day tankers, bulk carriers and cargo vessels are very large and can easily carry dangerous devices, substances and stowaways within spaces internal to the ship.

Ship Husbandry – Shipping companies often purchase old ships, hire crew for low wages and of low competence. Besides, shipping companies do not meet international ship safety standards. Adding to this problem is the desire on the part of shipping agents to transport cargo by the cheapest means, which ultimately encourages poor standards of maritime transportation. The maritime transport committee of the Organization for Economic Cooperation and Development (OECD), whose members include 30 of the world's leading states, has warned that substandard ships present risks of losses of life and environmental damage and urged governments, ship owners, shippers and the maritime industry to take immediate steps to stop unscrupulous operators from offering substandard ships for transporting cargo.

Illegal Activities – It will be unfair to entirely blame the shipping companies in this regard. The ship's crew has frequently engaged in activities such as drug smuggling, gun running and transporting human cargo. They also transport cargo that is not part of the legitimate consignment. For instance, on June 25 1999, a North Korean vessel, M V Kuwolsan, was impounded by Indian customs authorities at Kandla Port for carrying equipment for the production of tactical surface-to-surface missiles. The cargo, comprising 148 boxes, also included special materials and parts for guidance systems, blueprints, drawings and instruction manuals. Reportedly, the owners of the vessel admitted that the consignment was to be offloaded at Karachi Port in Pakistan. Then there is the nexus between corrupt port/customs officials and the crew that further threatens the safety of maritime transport. Shipping agents have been advised to beware of criminal "scams" to move large numbers of illegal immigrants around the world under the guise of being seafarers. The International Transport Intermediaries Club (ITIC) says it has received information on more than 30 cases where migrant smugglers approached shipping agents seeking to involve them in such activities.

Centre of Gravity – Disorder at Sea

In the recent past, piracy-related incidents, which were more common in the Malacca Strait and South China Sea, have tended to spill over into the Indian Ocean: Arabian Sea, Bay of Bengal and Horn of Africa. Similarly, the phenomenon of terrorism has been gathering momentum, and, among the two dozen or so terrorist groups identified to have been engaged in maritime terrorism, at least nine are currently active, and two of these operate in the Indian Ocean littoral.

According to military experts, future conflicts will take place in the littorals – that is, where sea meets the land (Shultz and Pfaltzgraff, 2000, 156). A large proportion of the world population is located in the littoral. Besides, much of the industrial infrastructure and wealth are concentrated in the areas. Coastal regions serve as the nodes for transport of trade, culture, and also as the hub of illegal activity, be it contraband trade, drug smuggling, gun running or even human smuggling. The sea serves as an easy highway and acts as a catalyst for promoting such activities. A quick look at the geography of the Indian Ocean region indicates that terrorist hubs are located in the littoral: LTTE in Jaffna, Sri Lanka; Al Qaieda in Yemen, Somalia, Pakistan and Indonesia, and the Free Aceh Movement in Indonesia. These hubs are also home to pirates. It can be argued that the center of gravity of piracy and terrorism at sea is currently located in the Indian Ocean region. This is further complicated by the Golden Crescent and the Golden Triangle hubs for drug smuggling and gun running.

Conclusion

With the passage of time, pirates and terrorists have sharpened their skills and also their tactics. These involve attacking ships both in harbour and at sea. They are known to use a variety of weapons, from knives to improvised explosive devices, submersibles, mini-submarines and high-speed boats. Dual-use technologies such as GPS, satellite communication systems, and water sport and scuba diving equipment are part of their inventories. At sea, terrorist groups have used rocket-propelled grenades, explosive-laden speedboats and even armour-piercing weapons. Their networks have kept pace with the changing technologies and they have adapted themselves to counter the strategies of maritime forces.

As noted, maritime order in the Indian Ocean is now being challenged from several directions. Piracy, terrorism and illegal activities at sea have the potential to disrupt sea-lanes. Efforts have been made by several countries to address this problem but only in terms of bilateral, and, at best, trilateral arrangements. There is a near total absence of a multilateral approach to safeguard sea-lanes in the Indian Ocean. Although multilateral Track II arrangements like the Council for Security Cooperation in Asia Pacific (CSCAP) and Western Pacific Naval Symposium (WPNS) have been actively engaged in hosting seminars and discussions, states are yet to incorporate their agenda into national and regional policy-making. This calls for greater cooperation among Indian Ocean regional states to build upon the recommendations of Track II arrangements.

Note

The views expressed in this paper are solely those of the author based on his on-going research and the information available through the ICC and IMB.

References

Arms seizures Backfirest < http://www.stratfor.com>.
Containerisation <http://www.choicegroup.co.in/html/cntrzation.htm.
Crew Scam Alert < http:// www.planetark.org>.
Gunaratna, R., "Trends in Maritime Terrorism - The Sri Lanka Case" *Lanka Outlook*, Autumn 1998,p.13.
Gunaratna, R., "Illicit transfer of Conventional Weapons: The Role of State and Non-state Actors in South Asia", in ed. Jayantha Dhanapala, Ambasssador Mitsuro Donowaki, Swadesh Rana, Lora Lumpe *Small Arms Control, Old Weapons, New Issues* (Aldershot: Ashgate, 1999), p.266.

Hanlon, E., "Taking the Long View: Littoral Warfare Challenges" in Richard H. Shultz Jr. and Robert L. Pfaltzgraff Jr., eds. *The Role of Naval Forces in 21st Century Operations*, p.156.

<http://www.pilotonline.com/military/ml0112sing.html>

ICC   PiracyReport<http://www.iccwbo.org/home/news_archives/2002 /piracy_report.asp>

IMB Calls For Clamp-Down on Fake Maritime Documents <http://www. iccwbo.org/index_ccs.asp>.

International Maritime Bureau Annual Reports and Quarterly Reports - various   years.   <   http://www.eia.doe.gov/emeu/cabs/choke. html#SUEZ>.

Maritime Security Measures Take Shape At IMO < http://www. imo. org/ Newsroom/mainframe.asp?topic_id=583&doc_id=2435>.

Menefee,

Murky Flag-Of-Convenience Ship Registry System Could Hamper Effort To Uncover   Terrorist   Assets   http://www.amo-union.org/Newspaper/ Morgue/11- 2001/Sections/News/foc.html

Knight Rider, US Ports Represent Weakness In Nation's Defence Analyses Show < http://www.military.com>

Killing Of Sea Bird Not A Big Blow to LTTE Shipping Operations, *The Sunday Times*, February 1996.

Operation   Cactus   <http://www.bharat-rakshak.com/CONFLICTS/ Operationcactus. html>.

Pallone raises issue of North Korean missile help to Pakistan
< http://www.indiainnewyork.com/iny0733099/chronicle/Pallone/html>

Piracy quarterly reports <http://home.wanadoo.nl/tortuga/ archive/modern/ modern.htm>.

Port of Entry Now Means Point of Anxiety
http://college4.nytimes.com/guests/articles/2001/12/23/892576.xml

Report on Ship attack in Sri Lanka, *FBIS*, 11 September 1997.

Review of Maritime Transport 2001 http://www.unctad.org/en/ press/ ms0105en.htm

Samuel Pyeatt Menefee, ' Piracy, Terrorism and Insurgent Passenger' in Natoline Ronzitte, 'Maritime Terrorism and International Law', (London : Maritnus Nijoff, 1991) ,pp.43 -55 .

'Stowaway Terrorists Steal into America by Sea Container' at DEBKA-*Net-Weekly Intelligence Report,* 18 June, 2002 <http://www.debka. com/LADEN/body_laden.html.>

Substandard Ships on Way Out < http://www.planetark.org>.

Terrorists wanted to hit US Frigate' at http://maritime.com/.

Terrorist In A Box: Business-class Suspect Caught In Container <http://hypocrisytoday.com/stowaway.html>.

Terrorism Probe Extends To Shipping', American Maritime Officer
http://www.amo-union.org/Newspaper/Morgue/10-
2001/Sections/News/foc.html

The World's Oceans Could Be The Next Target In The War On Terrorism <http://www.emergency.com/cntrterr.htm

Those Responsible Caught, <http://www.cargolaw.com/2000nightmare _cole.html.>

US Container Security Measures Take Shape, Internationally and Domestically http://www.tdctrade.com/alert/us0213.htm

What are FOCs ? < http://www.itf.org.uk/seafarer/foc/Body_foc..html.>

Worries Grow Over Tankers' Vulnerability to Attack' at http://www. planetark.org. Xinhua (Beijing),

CHAPTER 13

Freedom of Navigation and Indian Ocean Security: A Geopolitical Analysis

Sam Bateman

Introduction

The Indian Ocean is a major link in the global shipping network. Important shipping trades across the Indian Ocean include "Round the World" container services and oil routes from the Red Sea and the Persian Gulf to Asia via the Malacca or Lombok Straits and to Europe via the Suez Canal or the Cape of Good Hope. About half the world's seaborne trade in crude oil, an estimated 713 million tonnes in 1995, is shipped from the Middle East with about 14 per cent (99.6 million tonnes in 1995) bound for Europe, 12 per cent to North America (87.4 million tonnes) and 37 per cent (265 million tonnes) to Japan.[1] The safety and security of Indian Ocean shipping is a vital concern of many countries around the world.

Access to the Indian Ocean is either via critically important access waterways in the Northeast and the Northwest or from South of Australia or Africa. In the Northeast, access is via Singapore and Malacca Straits or several routes through the Indonesian archipelago or North of Australia. In the Northwest, it is via Bab el-Mandab strait connecting the Red Sea with the Gulf of Aden and the Arabian Sea or the Strait of Hormuz connecting the Persian Gulf with the Gulf of Oman and the Arabian Sea. These are all major "choke points" due to the high density of shipping traffic and their potential for closure either by the adjacent coastal states or by a maritime Superpower and/or a coalition of powerful naval forces.

They are invariably susceptible to mining and the focal areas in their approaches might offer attractive "hunting grounds" for submarines. The Mozambique Channel and the Maldive and Laccadive Islands create other "choke points" for shipping in the Indian Ocean. The area off Cape Agulhas, the most southerly point of Africa also constitutes a "choke point" for shipping transiting between the Indian Ocean and the South Atlantic.

Shipping using the Indian Ocean is vulnerable to numerous possible threats. These range from those induced by human activity such as piracy, maritime terrorism and human error leading to risks of collisions and groundings, particularly in focal areas and "choke points", to the natural hazards of bad weather, freak high waves and cyclones. Of the 47 bulk carriers that sank around the world between 1988 and 1991 apparently due to ship failure in severe sea conditions, 15 sank in the Indian Ocean with 112 of the 381 lives lost.[2] Another possible threat to the free movement of shipping might arise if a coastal state introduces strict regulations over shipping passing through its adjacent waters and maritime zones.

Acts of maritime terrorism in the Indian Ocean region include various attacks by the Tamil Tigers, the suicide boat attack on the USS *Cole* in Yemen in 2000 and most recently the attack on the French tanker *Limburg* off the coast of Yemen in early October 2002. This attack is now attributed to al Qaeda. Piracy also continues to be a major concern in South and Southeast Asian waters and off parts of Africa. The threat of a terrorist attack against a ship is of increasing concern to IMO and ship owners, particularly so in the case of vessels carrying dangerous cargoes, such as tankers, chemical tankers and LNG/LPG carriers.

The serious collision in the northern Malacca Strait in September 1992 involving the supertanker *Nagasaki Spirit* and the container ship *Ocean Blessing* may have been the result of a piratical attack on the latter vessel. However, this will never be proven as all onboard were killed in a fire and explosion after the collision.[3] This serious collision, as well as other marine casualties elsewhere in the Malacca and Singapore Straits, illustrates the need for effective traffic management in focal areas and "choke points". In a recent high profile marine casualty in the Indian Ocean, the Italian flag container vessel *Jolly Rubino* went aground on the Kwazulu Natal coast north of Durban and subsequently caught fire threatening environmental damage to the Greater St Lucia Wetland Park, South Africa's first World Heritage site. The cargo on the *Jolly Rubino* included toxic chemicals that had to be carefully removed from onboard in a long and dangerous operation.

Navigational Rights and Freedoms

In view of the high number of "choke points" in the Indian Ocean and the strategic importance of much shipping using the ocean, navigational rights and freedoms are an important issue for the region. The free movement of shipping is a common interest both of littoral and island countries as well as other countries in Asia-Pacific and Europe that are also dependent on Indian Ocean shipping. Depending on the actual route chosen, ships transit either on the "high seas" or through waters under some form of jurisdiction of one coastal state or another. Different navigational rights and obligations apply to the various zones of coastal state jurisdiction. The main passage regimes that apply are discussed in this section along with possible problems that might arise.

Innocent Passage

Under the 1982 UN Convention on the Law of the Sea (LOSC), ships of all states enjoy the right of innocent passage through territorial seas of other states.[4] Innocent passage is the most restrictive of the passage regimes and may be suspended in certain circumstances,[5] submarines must travel on the surface and show their flag[6] and ships are prevented *inter alia* from operating organic aircraft.[7] Many countries still regard the obligation to allow foreign ships the right of innocent passage through their territorial sea as a significant limitation on their sovereignty over and a potential threat to their national security.

A major problem is the requirement of some coastal states for prior notification or authorization for the innocent passage of certain vessels, particularly nuclear-powered ones and warships. There are over 40 states around the world that place restrictions on warship innocent passage. These include the following Indian Ocean countries: Bangladesh, Burma, India, Indonesia, Iran, Maldives, Mauritius, Oman, Pakistan, Seychelles, Sri Lanka and Yemen.[8] Arguments against prior authorization or notification are supported by the failure at the Third UN Conference on the Law of the Sea (UNCLOS III) to have the requirement explicitly included in the Convention.[9] The U.S. sees such a requirement as incompatible with the freedom of navigation and the spirit of the Convention.[10] Arguments for prior authorization have less justification than those for prior notification, as they would be specifically contrary to LOSC Article 24 that prohibits coastal state regulation that "hampers, denies, or impairs the right of innocent passage".[11] However, the arguments for prior notification have some merit not necessarily in legal terms, but at least politically. For example, secrecy of movement may be seen as incompatible with the processes of maritime confidence and security building and transparency.

Straits' Transit Passage

The LOSC Part III established a special regime to apply when a strait used for international navigation is wholly or in part contained within the territorial sea of one or more states. The straits in the Indian Ocean region where this regime applies include Bab el-Mandab, the Straits of Hormuz and the Singapore and Malacca Straits. In effect this regime establishes a new maritime zone for the waters within an international strait although there is no precise means of establishing the geographical limits of the "zone". The straits' transit passage regime applies to "straits which are used for international navigation between one part of the high seas or an exclusive economic zone and another part of the high seas or an exclusive economic zone",[12] or in order to enter or leave a state bordering the strait.[13] It is available to all ships and aircraft.[14] While transit passage may not be suspended,[15] transiting ships must comply with generally accepted international regulations, procedures, and practices for safety at sea and for the prevention, reduction, and control of pollution from ships.[16] Also, the states bordering these straits may introduce appropriate regulations although these must be non-discriminatory in their application.[17]

Maritime jurisdiction associated with transit passage is complex and problems could arise in the future if coastal states extend controls over shipping beyond those regarded as acceptable by the major maritime powers. Ship exercising the right of transit passage pass through several different jurisdictional zones possibly under the jurisdiction of different coastal states. By definition, the strait contains territorial waters and then to enter and leave the strait, the ship must pass through the EEZ of one or more coastal states.

Numerous legal and practical problems can be identified with the application of the transit passage regime.[18] A particular problem relates to the "extent" of a strait and, consequently of the area where the transit regime applies?[19] Does transit passage only begin when the vessel crosses the 24-mile closing line, or does it begin at some earlier stage of the transit?[20] How close may the vessel approach the coast of a bordering state while still exercising the freedom of transit passage? It is unclear when transit passage begins. The LOSC does not define the extent of a strait used for international navigation and "It follows that the area in which transit passage rights can be claimed is therefore left perhaps deliberately vague".[21] Unlike with archipelagic sea lanes (ASL) passage discussed below, there are no limits placed on choice of route through the strait. Unless sea lanes or traffic separation schemes are designated in accordance with LOSC Article 22, vessels are allowed to enter or leave an international strait at any point on the 24-mile closing line, and similarly enter or leave the strait by any route. Hence transit passage cannot be confined to the area of water where an international strait is entirely enclosed by the adjacent territorial sea.[22]

Coastal states adjoining an international strait have considerable service responsibilities for vessels transiting the strait (e.g. navigational aids, hydrographic charts and other navigational nformation, search and rescue services, and marine pollution contingency arrangements) but LOSC makes no provision whatsoever regarding any form of cost-recovery.[23] These considerations led a Malaysian maritime expert to refer to the transit passage regime as "fundamentally flawed" because it puts the entire burden of managing the straits on the coastal states.[24] States which have sought to place restrictions on the transit passage regime include Iran, Oman, Spain and Yemen with the U.S. protesting all these claims and conducting operational assertions of rights.[25]

Archipelagic Sea Lanes (ASL) Passage

The LOSC Part IV established the regime of the archipelagic state, which allows countries constituted wholly of one or more groups of islands and meet certain other criteria to draw archipelagic baselines joining the outermost islands and drying reefs. The Maldives and Indonesia are archipelagic states in the Indian Ocean region. Archipelagic states exercise full sovereignty over their archipelagic waters qualified only by the regime of ASL passage which allows ships of all states the right of "continuous, expeditious and unobstructed transit" through archipelagic waters along sea lanes which may be designated by the archipelagic state.[26] If sea lanes are not designated, then the right of ASL passage may be exercised through the routes normally used for international navigation.[27] Outside ASLs, ships of all states have the right of innocent passage only and must abide by the more restrictive provisions of that regime, including recognition of the principle that the archipelagic state may temporarily suspend innocent passage.[28]

Archipelagic states have been firmly of the view that the ASL passage regime applies in their straits into or within their archipelago rather than the straits transit regime. The right of ASL passage and overflight is restricted to particular routes and a particular route axis (with no deviation more than 25 miles from the axis line and a restriction on navigating closer to the coasts than 10 per cent of the distance between the nearest points on islands bordering the sea lane).[29] As was noted earlier, there is no similar requirement for ships and aircraft on transit passage to use "normal" routes, to keep away from the coast, or to keep their passage within a certain distance of an axis line.

Exclusive Economic Zone

The Exclusive Economic Zone (EEZ) regime established under the LOSC Part V grants coastal states rights over the living and non-living resources of the 200 nautical mile EEZ but also imposes certain obligations on them to preserve and protect the marine environment.[30] Otherwise, the high seas freedoms of navigation are considered to prevail although LOSC Article 220 grants certain enforcement powers to coastal states in respect of the prevention, reduction and control of pollution of vessels. This could lead to laws and regulations that infringe upon navigational rights and freedoms,[31] and the United States, for example, has introduced legislation in its Oil Pollution Act 1990 (OPA 90) to restrict tanker operations in its EEZ.[32]

The EEZ regime was a major development in international law, the full implications of which are only now being appreciated. An important area of disagreement relates to the ability of a coastal state to introduce regulations that have the effect of denying freedoms of navigation and overflight in all or part of its EEZ. The maritime powers argue that, subject to the resource-related rights and environmental protection obligations of a coastal state, the freedoms of navigation and overflight in the EEZ are the same as those on the high seas.[33] This is in accordance with LOSC Article 58, which states that in the EEZ, all states enjoy those freedoms referred to in Article 87 "of navigation and overflight and of the laying of submarine cables, and other internationally lawful uses of the sea related to those freedoms, such as those associated with *the operation of ships, aircraft* and submarine cables and pipelines, and compatible *with the other provisions of this Convention*" (emphasis added). The position of the maritime powers on this issue gains some support from the fact that the proposals at UNCLOS III to restrict the holding of military exercises in the EEZ were unsuccessful.[34] However contrary to these arguments, some coastal states have declared security zones that extend into the EEZ, or have specifically claimed that other states are not authorised to conduct military exercises or manoeuvres in the EEZ without their consent.[35] Indian Ocean countries that apply these restrictions include Bangladesh, India, Malaysia, Myanmar, Pakistan, Sri Lanka and Yemen.

Focal Areas and "Choke Points"

Northwest Indian Ocean

Bab el-Mandab – The Bab el-Mandab is the narrow entrance into the Red Sea from the Gulf of Aden. The riparian states are Djibouti and Eritrea to the West and Yemen to the East. The strait is about 14.5 nautical miles wide at the narrowest part of the passage. Oil traffic passes through the strait in both directions with loaded tankers from ports in Saudi Arabia in the Red Sea heading towards East Asia and to the Cape of Good Hope and ones from ports in the Gulf heading towards the Mediterranean and Western Europe through the Suez Canal. The Suez Canal at the northern end of the Red Sea has a depth limitation of 58 feet and Panamax-size vessels[36] are the largest vessels that can use the Suez Canal. Very large crude carriers (VLCCs) and ultra large crude carriers (ULCCs)[37] cannot use the Suez Canal and such vessels loading in Saudi ports must use the Bab el-Mandab and transit around the Cape of Good Hope if heading for North America or Western Europe. The Egyptian Government plans to widen and deepen the Suez Canal so that by 2010 it will be able to accommodate VLCCs and ULCCs.[38]

Total oil traffic through the Bab el-Mandab was estimated to be 3.2-3.3 million barrels per day in 2000.[39] Closure of the strait would keep tankers from the Persian Gulf from reaching the Red Sea and divert them around the Cape of Good Hope adding greatly to transit time and cost, and effectively tie up spare tanker capacity. The Bab el-Mandab could be by-passed by northbound traffic by utilizing the East-West oil pipeline that traverses Saudi Arabia. However, southbound oil traffic would still be blocked. Closure of the strait would also effectively block all non-oil shipping from using the Suez Canal. Oil tankers constitute about one-fifth of all shipping traffic through the Suez Canal.[40]

Strait of Hormuz – The Strait of Hormuz connects the Persian Gulf with the Gulf of Oman and the Arabian Sea and is bordered to the West by Oman and to the East by Iran. Although the area of overlap of twelve nautical mile territorial seas between the two riparian countries is relatively small, the regime of straits' transit passage applies but the interpretation of that regime in the strait has been the subject of diplomatic exchanges between the U.S. and Iran.[41] The strait comprises two-mile wide channels for inbound and outbound tanker traffic, as well as a two-mile wide buffer zone. It is by far the world's most important oil choke point with an oil flow estimated in 2000 to be 15.5 million barrels per day.[42]

Closure of the Strait of Hormuz would have huge strategic impact on Northeast Asia, North America and Western Europe and it is not surprising that the U.S. has attached such importance to the freedom of navigation through the strait including basing warships at Bahrein inside the Gulf. There is really no alternative to tankers for getting oil to East Asia and only a fairly complicated alternative for North America using a combination of tankers and pipeline that would involve increased transportation costs.

Maldives and Dondra Head – East-West shipping transiting between the Red Sea or the Gulf and Southeast Asia must pass through the barrier of islands stretching over 700 nautical miles from roughly 12 degrees North to the equator and comprising the Laccadive Islands under Indian sovereignty and the Republic of Maldives. The main options are the Nine Degree Channel (within the Indian EEZ), the preferred Eight Degree Channel (partly within the EEZ of India and that of the Maidives) and the One and a Half Degree Channel (within the EEZ and archipelagic waters of the Maldives). The actual choice of passage would depend on the strait to be used subsequently for transiting Southeast Asian waters. From October to April and to take account of the prevailing Northeast monsoon, ships using Sunda or Lombok Straits will likely prefer the One and a Half Degree Channel.[43]

The Maldives declared an EEZ in 1986. This was rectangular in shape and based on a "picture frame" claim to territorial sea. This latter claim was unacceptable under modern international law and the Maldives has moved to claim status as an archipelagic state under the LOSC. Base points have been established for archipelagic baselines. Relevant domestic legislation was passed in 1996. No action has been taken so far to declare ASLs through the Maldivian archipelago but these would seem to be a requirement (particularly through the One and a Half Degree Channel, Kardiva Channel and the Equatorial Channel). After clearing the Laccadive/Maldives island chain, shipping then passes through the EEZ of Sri Lanka. If heading towards the northern entrance to the Malacca Straits, ships may also pass through the territorial sea of Sri Lanka exercising the right of innocent passage. If heading toward Sunda or Lombok Straits, they may be further offshore. The focal area off Dondra Head, the Southern extremity of Sri Lanka, constitutes a significant "choke point" for shipping transiting between the Suez Canal or the Gulf to the Malacca Strait.

Northeast Indian Ocean

The area between Singapore and Darwin in northern Australia and the straits between the Indian and Pacific Oceans through and adjacent to the Indonesian archipelago now constitute the most significant shipping "bottleneck" in the world. The vessels passing through this area carry the sources of energy (oil, LNG and LPG) and raw materials essential for the economic growth and survivability of China,[44] Japan, South Korea and Taiwan, as well the large container ships on the main around the world route linking Europe to East Asia. The value of this trade totaled nearly a trillion US dollars in 1994,[45] and it required over half the total shipping capacity of the world to move it.[46]

Malacca and Singapore Straits – This "bottleneck" between Singapore and Darwin is most confined in the Malacca and Singapore Straits where, at its narrowest point in the Phillips Channel of Singapore Strait, the passage is a bare 2.5 nautical miles wide.[47] These straits provide the most direct route through the "bottleneck" and are regarded as the key shipping choke point in Asia. If the straits were closed, nearly half of the world's fleet would be required to sail further generating a substantial increase in global demand for shipping. More than 50,000 vessels per year transit the straits per year. India has demonstrated its interest in the safety and security of shipping through the straits by assisting the U.S. with the escort of selected high value ships through the strait against the risk of piratical attacks, which are prevalent in the area, and terrorism. Similarly Japan contributes financially to the maintenance of navigational aids and hydrographic surveys in the Malacca Strait and in recent years has also been providing ships and aircraft from the Japan Coast Guard to assist in anti-piracy measures in Southeast Asia.

There is a least depth in the straits of about 25m and ships require an under keel clearance of 3.5m in accordance with the International Maritime Organization's (IMO's) Rules for Vessels Navigating through the Straits of Malacca and Singapore.[48] Very large oil tankers over about 250,000dwt will probably be outside these draft limitations. These ships normally use Lombok and Makassar Straits further Eastward although this might add about one thousand nautical miles and three days' steaming to the passage.[49]

The states bordeiing the Malacca Strait have from time to time contemplated compulsory pilotage schemes as part of their ability to control certain aspects of navigation that could impact upon the marine environment.[50] However, such schemes have not been introduced because refusing access to a strait to a vessel on the grounds that it would not accept a pilot would amount to hampering transit passage and be contrary to LOSC Article 44 in particular.[51] Indonesia may prefer the compulsory re-routing of larger tankers through the Lombok Straits.[52] Both Indonesia and Singapore backed Malaysia's insistence that Japanese plutonium shipments should not be routed through the Malacca Strait.[53]

The IMO has introduced a mandatory ship reporting scheme for the Malacca and Singapore Straits,[54] and Indonesia, Malaysia, Singapore and the IMO have also agreed to go ahead with the establishment of a Marine Electronic Highway for the Malacca and Singapore Straits.[55] This integrated system will include electronic nautical charts, positioning systems, automatic ship identification (AIS) transponders, as well as the provision of meteorological, oceanographic and navigational information. It will provide an essential tool for marine pollution prevention, marine pollution control, marine environmental planning and management, as well as safety of navigation. It will allow for the maximum of information to be made available to ships as well as shore-based users such as the vessel traffic control systems managed by the coastal states adjacent to the straits.

Sunda and Lombok Straits Indonesia's proposal in the early 1990s to designate three North/South ASLs subsequently led to detailed analysis and discussion at the IMO and eventual approval in 1998 of the "General Provisions on the Adoption, Designation and Substitution of Archipelagic Sea Lanes" (GPASLs).[56] The three ASLs adopted were:

- Sunda Straits – Karimata Strait in the Western part of the archipelago;
- Lombok Strait – Makassar Strait in the Central part; and
- Banda Sea – Moluccan Sea in the Eastern part.

Sunda Strait between Sumatra and Java offers the most direct route from the Indian Ocean into the Java Sea but the least depth in this strait is around 27 metres and the area is relatively poorly charted. It is only an alternative for ships that could use the Malacca and Singapore Straits in any case. Deeper draught vessels must use Lombok Strait between the islands of Bali and Lombok and then proceed northwards by the Makassar Strait East of Borneo. The Lombok and Makassar Straits are both wide, deep and navigationally straightforward. In 1988 Indonesia reportedly closed the Sunda and Lombok Straits for a period of time leading to strong diplomatic protects from the U.S., Australia and other countries although the circumstances of the closure and Indonesia's intentions with the act remain unclear.[57]

Political Factors and Problem Areas

The Need for Cooperation and Confidence-Building

The concept of the freedom of navigation has both economic and strategic significance. Disruption of the free flow of commercial shipping by the most expeditious route can have serious economic repercussions leading to higher transport costs, delays in supply of critical imports and increased demand for shipping capacity. More serious however, are the strategic repercussions. The disruption of Indian Ocean seaborne trade could deprive some countries, including Japan, China and to a lesser extent, the U.S., of essential imports, particularly of energy. Furthermore the access waterways into the Indian Ocean have great strategic significance for

Indian Ocean countries wishing to deploy naval forces out of the Indian Ocean, and for extra-regional countries wishing to bring their naval forces in.

The current dependency of Northeast Asia on oil imports from the Middle East of 70 per cent has been predicted to rise to 95 per cent by 2010 and:

> Should this projected pattern indeed materialize, a growing fleet of heavily laden supertankers will plow east across the Arabian Sea and the Indian Ocean in coming decades, headed for Singapore, Hong Kong, Shanghai, Pusan, and Yokohama. East Asian oil imports from the Middle East could well triple in the next fifteen years alone, to a very substantial share of total world oil consumption.[58]

As these trends continue, countries may be expected to commit increased political and diplomatic effort into ensuring the freedom of navigation in the Indian Ocean. Fundamentally this should mean cooperation rather than unilateral actions. Major Indian Ocean countries, particularly Australia, India and South Africa, share a common interest with important non-Indian Ocean countries in Northeast Asia, North America and Western Europe in maintaining freedom of navigation in the Indian Ocean, especially the key access waterways. This common interest establishes an essential basis for maritime cooperation and confidence building in the Indian Ocean.

Cooperation is required to enhance the commercial efficiency of shipping operations, maintain safety and reduce the impact of these operations on the marine environment including by the reduction of the number of shipping casualties. Cooperation is also essential for the mitigation and prevention of ship-sourced pollution, including arrangements for responding to the major oil spills that may result from maritime accidents and monitoring oil spills in open waters from routine tank cleaning and ballasting operations. It is also required in an operational context to prevent and deter piratical and terrorist attacks. In a period of tension or conflict, operational cooperation would also be necessary to protect focal areas and choke points against closure due to mines or other threats to shipping.

Following the 11 September 2001 terrorist attacks in New York, IMO has given high priority to the review of existing international legal and technical measures to prevent and suppress terrorist attacks against ships and improve security aboard and ashore. The aim is to reduce the risk to passengers, crews and port personnel both onboard ships and in port areas and to vessels and their cargoes. Measures being considered include accelerated introduction of a mandatory system of automatic identification (AIS) fitted to all ships over 500 gross tones on international voyages; updated arrangements for seafarer identification; and the introduction of a proposed International Ship and Port Facility Security Code (ISPS Code). The latter code would include requirements for ships and port facilities such as the provision of security plans and officers, onboard equipment and arrangements for monitoring and controlling access. Regional implementation of new IMO measures will be important although they place additional burdens on developing countries.

Port State Control

One positive development for marine safety in the Indian Ocean that has occurred recently is the establishment of the Indian Ocean Memorandum of Understanding (MOU) on Port State Control (PSC) with a secretariat in Goa in India. PSC involves the inspection by the officers of a national maritime authority of foreign vessels entering port to ensure compliance with international maritime safety and marine pollution conventions. PSC supplements the primary responsibility of the Flag State for ensuring the compliance of ships flying its flag with the relevant international conventions. The Indian Ocean MOU provides for cooperation on ship inspections with a regional database and agreement on target inspection rates for foreign ships visiting the ports of participating countries. At present these include Australia, Bangladesh, Djibouti, Eritrea, India, Iran, Kenya, Maldives, Mauritius, Mozambique, Myanmar, Seychelles, South Africa, Sri Lanka, Sudan, Tanzania and Yemen.[59]

Navigational Regimes

A common regional understanding of aspects of navigational regimes where uncertainty exists, including coastal state rights in the EEZ and aspects of the transit passage regime, would constitute an important confidence building measure.[60] While differences on navigational issues do not usually cause problems, they can become dangerous when tensions exist, and any measures at all that would have the effect of limiting the scope for disputation would be advantageous. The general trend towards a divergence of state practice with regard to the rights of coastal states and other states in the EEZ is an unhelpful development that could impact on the freedom of navigation in the future.

The CSCAP Maritime Cooperation Working Group established by the Council for Security Cooperation in the Asia Pacific (CSCAP) has developed a CSCAP Memorandum on the Practice of the Law of the Sea issues in the Asia Pacific. This should be released in early 2003 although due to the differing views of CSCAP members it was unable to produce a definitive statement on EEZ rights. Another forum that is investigating specifically the rights of States to conduct military and intelligence gathering activities in the EEZs of another coastal State is the series of policy and planning workshops organized by the East-West Center in Honolulu.[61]

The law of the sea is facing a need to adjust to the possible implications of the Information Technology (IT) revolution, particularly on navigational regimes and the ability of a coastal State to track the movement of vessels in its adjacent waters. The IT revolution and associated technologies, particularly satellite monitoring systems, provide coastal States with the ability to track the position of any ship in its territorial sea, in its EEZ, or above its continental shelf. However, there are implications for the freedom of navigation, as well as for the principle of sovereign immunity.[62] A coastal State could insist that warships activate their locator devices (transponders) while passing through the State's territorial sea, including in an international strait.[63] Some countries might argue that compulsory fitting of AIS transponders is an interference with the freedom of navigation. The U.S. and other states might argue that, even if the transponders were fitted to all commercial vessels, the principle of sovereign immunity would apply to warships and other government vessels.

Barriers to Cooperation

The Indian Ocean presents some peculiarly tricky problems for security management, including ensuring the freedom of navigation. Despite their common interest in the freedom of navigation, and for that matter oceans management generally, the littoral and island countries of the Indian Ocean are extremely diverse. Previous attempts at building cooperation in the Indian Ocean region have not been resounding successes and the level of existing cooperation is not as high as elsewhere in the world. The Indian Ocean rim is the focus of most of the world's underdevelopment, population growth, the issues that produce North-South schisms and frustrations, disease and natural disasters (including tsunamis and tropical storms) and environmental vulnerability. There is also the issue of the extent to which non-Indian Ocean countries should be involved in cooperation although globalization has tended to weaken the objections to their involvement. Certainly this is the case with cooperation to ensure the freedom of navigation.

Despite the clear benefits of cooperation, there are still many barriers to cooperative activities. It is not hard to find problems with strengthening maritime security cooperation. Countries have suspicions concerning the capabilities and intentions of their neighbours. There are few obvious naval "partners" in the Indian Ocean. Problems exist with common doctrine, language and interoperability of equipment. Regional navies acquire their ships, submarines and aircraft from a wide range of sources. The problems involved become even more acute as the technological levels of navies increase. Technical deficiencies in some navies may significantly inhibit cooperation with less advanced navies being reluctant to engage in operational cooperation for fear that their deficiencies will be too apparent.

Cooperative activities may be used to gain intelligence on the capabilities of another country. Even innocuous naval port visits can provide an opportunity to gather intelligence both by the host country collecting information about visiting ships and by visiting ships about the host state. This might include signals intelligence gained by the foreign warships listening in on the host country's naval communications while they are in port. Expert intelligence collectors can obtain much vital information on another navy, particularly data on weapons, sensors and communications systems (including the possible identification of highly sensitive frequencies to support prospective electronic warfare), during operations with ships and aircraft of another country.

Conclusions

The freedom of navigation is extremely important in the Indian Ocean due to the strategic and economic importance of shipping using the ocean and the number of major choke points. It will become even more important as the process of globalization continues. The freedom of navigation may be disrupted by physical actions including acts or piracy or terrorism or by military action in time of conflict. It may also be disrupted by coastal states, particularly ones adjacent to choke points, seeking to impose stricter regulations on passing ships than those generally accepted by the international maritime community. All countries, both coastal and user states, have a common interest in the provision of services to ships that will improve the safety of navigation and reduce the incidence of marine casualties in the Indian Ocean, particularly in the choke points. IT developments, such as AIS and the Marine Electronic Highway, facilitate these services.

There is scope for a higher level of naval and maritime cooperation to ensure and enhance the freedom of navigation in the Indian Ocean. However, political care is required lest cooperative activities become "stumbling blocks" rather than "building blocks" for a more secure Indian Ocean. Activities that might be undertaken include the development of regional protocols to cover applicable navigational regimes and operational cooperation to protect shipping from piratical and terrorist attacks, as well as contingency planning to protect focal areas against threats in time of conflict. Development of these initiatives is not just a matter for Indian Ocean countries but requires the involvement also of major user states in Northeast Asia, Western Europe and North America that have an interest in the freedom of navigation in the Indian Ocean. These more developed countries have a role in building the capacity of less developed countries in the region to implement and manage navigational safety services. This is all a part of the processes of globalization.

Notes

1. Norges Rederiforbund, *World Seaborne Trade 1995*, http:/www. rederi. no/en/library/qi/qi/4-97/3e.html.
2. E.A. Bryant and W.S.G. Bateman, "Maritime Hazards in the Indian Ocean Region: An Overview" in Colin D. Woodroffe (ed), *Maritime Natural Hazards in the Indian Ocean Region*, Wollongong Papers on Maritime Policy No.6, Centre for Maritime Policy, University of Woolongong, 1998, p.26.
3. Peter Chalk, "Contemporary Maritime Piracy in Southeast Asia", Studies in Conflict, and Terrorism, Vol. 21, No. 1, January-March 1998, p. 97.
4. LOSC Article 17.
5. LOSC Article 25(3).
6. LOSC Article 20.
7. LOSC Article 19(2)(e).
8. J. A. Roach and R.W. Smith, *Excessive Maritime Claims*, International Law Studies Vo1.66, Newport R.I., USN War College, 1994, Table 10 pp. 158-159,
9. David F. Froman, "Uncharted Waters: Non-Innocent Passage of Warships in the Territorial Sea" San Diego Law Review, Vol. 21, No. 3, June 1984, p.642; and Roach and Smith, *Excessive Maritime Claims*,. p. 156.
10. U.S. Statement to UNCLOS III quoted in Roach and Smith, *Excessive Maritime Claims*, p. 156.
11. Froman, "Uncharted Waters", p. 654.
12. LOSC Article 37.
13. LOSC Article 38 (2)
14. LOSC Article 38 (1).
15. LOSC Article 44
16. LOSC Article 39(2).
17. LOSC Article 42.
18. Discussed in more detail in Sam Bateman, "The Regime of Straits Transit Passage in the Asia Pacific: Political and Strategic Issues" in Donald R Rothwell and Sam Bateman (eds), *Navigational Rights and Freedoms and the New Law of the Sea*, The Hague, Martinus Nijhoff Publishers, 2000, pp. 94-109.

19. Ibid., pp. 85-86.
20. Thos is a particularly important consideration for submarine navigation as in areas of territorial sea where the innocent passage regime only applies, submarines are required to travel on the surface and show their flag. Any suggestion that the transit passage regime did not apply in the territorial sea leading to the international strait would compromise the strategic element of submarine navigation. R.W. Smith and J.A. Roach "Navigation Rights and Responsibilities in International Straits: A Focus on the Straits of Malacca" MIMA Issue Paper, Kuala Lumpur, Malaysian Institute of Maritime Affairs, 1995, p. 8, and in B.A. Hamzah, The Straits of Malacca: International Cooperation in Trade, Funding and Navigational Safety. Kuala Lumpur, Pelanduk, 1997, p. 289.
21. J.B.R.L. Langdon, "The Extent of Transit Passage – Some Practical Anomalies", Marine Policy, Vol. 14, No. 2, 1990, p. 130.
22. Ibid., p. 135.
23. Roach and Smith, Excessive Maritime Claims, p. 7.
24. B.A. Hamzah, "The Security of Sealanes: The Search for an Equitable Straits Regime", Paper for Eighth International Conference on Sea Lanes of Communication (SLOC), Bali, 24-27 January, 1993, pp. 17-18. Alternative methods of covering the costs are explored in various papers in Hamzah, The Straits of Malacca.
25. Hamzah, "The Security of Sealanes", p. 17.
26. LOSC Article 53 (1), (2) and (3).
27. LOSC Article 53 (12).
28. LOSC Article 52 (2).
29. LOSC Article 53 (5).
30. LOSC Article 52 (2).
31. D.R. Rothwell, "Navigational Rights and Freedoms in the Asia Pacific Following Entry Into Force of the Law of the Sea Convention". Virginia Journal of International Law, Vol. 35, No. 3, Spring 1995, p. 598.
32. For a discussion of OPA 90 see John E. Noyes, "The Oil Pollution Act of 1990", International Journal of Estuarine and Coastal Law, Vol. 7, 1992, pp. 43-56; and Thomas J. Wagner, "The Oil Pollution Act of 1990: an analysis", Journal of Maritime Law and Commerce, Vol. 21, 1990, pp. 569-587.
33. George R. Galdorisi and Kevin R. Vienna, Beyond the Law of the Sea – New Directions for U.S. Oceans Policy, Westport, Praeger, 1997, p.151.
34. David James Attard, The Exclusive Economic Zone in International Law, Oxford, University Press, 1987, pp. 84-85.
35. R.R. Churchill and A.V. Lowe, The Law of the Sea, 3rd ed., Manchester, Juris Publishing, Manchester University Press, 1999, p. 427; and Ross, "Naval Activity in the Exclusive Economic Zone", p. 127.
36. Panamax is a reference to the maximum size of ship (container ship. bulk carrier, oil tanker or passenger ship) that can use the Panama

Canal. Typically they are up to about 80,000dwt with a maximum beam of 32m and a draft of 12 metres.

37. Very large crude carriers (VLCCs) are between 150,000 and 300,000 dwt and ultra large crude carriers (ULCCs) are over 300,000 dwt.

38. U.S. *Energy Information Administration*, World Oil Transit Chokepoints, November 2001, http://www.eia.doe.gov/emeu/cabs/choke.html

39. Ibid.

40. Of a total of 9969 ships that used the Suez Canal during the period January-September 2002, 1878 were oil tankers. Suez Canal Traffic, http://www.economic.idsc.gov.eg/suez.asp

41. Roach and Smith, *Excessive Maritime Claims*, pp. 189-191.

42. Ibid.

43. Captain Russ Swinnerton RAN, "A Description of Regional Shipping Routes: Navigational and Operational Considerations", *Maritime Studies 87*, March/April 1996, p.15.

44. A Chinese view of the strategic importance of shipping may be found in Ji Guoxing, "SLOC Security in the Asia Pacific", *Modern Management Research paper*, July I 999, Shanghai, Modern Management Researeh Centre, 1999.

45. H.J. Kenny, *An Analysis of Possible Threats to Shipping in Key Southeast Asian Sea Lanes*, CNA Occasional Paper, Alexandra, VA, Center for Naval Analyses, February 1996, p.16.

46. J.H. Noer with D. Gregory, *Chokepoints: Maritime Economic Concerns in Southeast Asia*, Washington DC, National Defence University Press, 1996, p. 8.

47. Vivian L. Forbes, "Ensuring Safety to Navigation in the Straits of Singapore", *The Indian Ocean Review*, Vol. 10, No. 3, September 1997, p. 7.

48. Swinneiton, "A Description of Regional Shipping Routes", pp. 10-22.

49. Ibid., p. 18.

50. Rothwell, "Navigational Rights and Freedoms", pp. 604-607.

51. Stuart Kaye, *The Torres Strait*, The Hague, Martinus Nijhoff, 1996, p. 85.

52. Swinnerton, "A Description of Regional Shipping Routes", pp. 10-22.

53. Jon M. Van Dyke "Sea Shipment of Japanese Plutonium under International Law", *Ocean Development and International Law*, Vol. 24, 1993, pp. 399-403.

54. Resolution MSC 73(69) adopted by IMO on 29 May 1998.

55. IMO Newsroom, "First phase of East Asia's Marine Electronic Highway takes off", http://www.imo org/Newsroorn 24 March 2001.

56. Robin Warner, "Implementing the Archipelagic Regime in the International Maritime Organization" in Rothwell and Bateman, *Navigational Rights and Freedoms*, p. 170.

57. D.R. Rothwell, "The Indonesian Straits Incident: Transit or archipelagic sea lanes passage", *Marine Policy*, Vol. 14, No. 6, November 1990, pp.491-506.

58. Kent E Calder, *Asia's Deadly Triangle – How Arms, Energy and Growth Threaten to Destabilize Asia Pacific*, London Nicholas Brealey Publishing, I 9%, p. 59.

59. The Tokyo MOU has a website at; http://www.iomou.org/

60. Sam Bateman, "Maritime Confidence and Security Building Measures in the Asian Pacific Region and the Law of the Sea" in James Crawford and Donald R Rothwell (eds) *The Law of the Sea in the Asian Pacific Region*, Dordrecht, Mattinus Nijhooff, 1995, p. 233.

61. The first of these workshops was held in Bali in June 2002 and the next in Tokyo in February 2003. The report of the Bali Dialogue is available at: <http://www.EastWestCenter.org/stored/pdfs/ Bali Dialogue.pdf>

62. Anthony Bergin, "Maritime Surveillance, Enforcement and Information Exchange" in Dalchoong Kim, Seo-Hang Lee and Jin-Hyun Paik, *Maritime Security and Cooperation in the Asia-Pacific towards the 21st Century*, East and West Studies Series 46, Seoul, Institute of East and West Studies, Yonsei University, 2000, pp. 143166.

63. John A. Knauss and Lewis M. Alexander, "The Ability and Right of Coastal States to Monitor Ship Movement: A Note", *Ocean Development and International Law*, Vol. 31, 2000, p. 379.

CHAPTER 14

The Future for Indian Ocean Cooperation

Sanjay Chaturvedi and Dennis Rumley

Our key intention in this concluding chapter is two-fold. First, to identify and elucidate, on the basis of our understanding of the individual as well as the collective thrust of the various contributions to this volume, some of the key issues which we believe have a long-term bearing on the prospects for realizing maritime security and cooperation in the Indian Ocean. Second, to pinpoint areas/sectors, where sub-regional and regional cooperation is desirable as well as feasible, and to examine the efficacy of the current mandate of the Indian Ocean Rim Association for Regional Cooperation (IOR-ARC) for achieving sustainability, stability and security in the Indian Ocean.

First and foremost, what emerges, rather forcefully, out of various essays in this collection, at a larger and deeper level, is that the Indian Ocean needs to be re-imagined as encompassing not only 'spaces of places' but also 'spaces of flows'. The post-Cold War Indian Ocean is characterized by a set of geopolitical equations and flows that cannot be captured and explained by classical geopolitical doctrines. In other words, the Indian Ocean region needs to be liberated from the prison of old cartography, a cartography that cannot map the globe girdling networks of flows in the form of corporations, trade and communications infrastructure. Such a perspective challenges conventional geographical scale and requires foreign policy, military and commercial establishments to reorient themselves. With both state and non-state actors looking for definitions and certainties in a period that is itself struggling to find answers, the issues related to identity, boundaries and sovereignty are best handled by the intellectuals and institutions of statecraft within a much larger oceanic vision. As E. M. Borgese (1998:6) has put it so succinctly:

> The ocean is a medium different from the earth: so different, in fact, that it forces us to think differently. The medium itself, where everything flows and everything is interconnected, forces us to "unfocus", to shed our old concepts and paradigms, and to "refocus" on a new paradigm. Fundamental concepts, evolved over the millennia on land, like sovereignty, geographic boundaries, or ownership, simply will not work in the ocean medium where new political, legal, and economic concepts are emerging. Eventually they will act on the social, economic and political order of the next century.

For four decades and more, the power-political intrigues and intricacies of the Cold War ideological geopolitics transformed the political and socio-cultural geographies of the Indian Ocean into a militarized geography, disallowed international cooperation among the Indian Ocean states in the civil domain and caused extensive damage to the terrestrial as well as marine ecosystems. It is fairly recent that the discourse of ocean governance, especially as it relates to ecologically sustainable and equitable development and management of fast multiplying uses of ocean space and resources, has caught the attention of academia and policy-makers concerned with the Indian Ocean. With 1998 designated by the United Nations as the "Year of the Oceans", and the attention of the media and the NGOs much more focused now on marine conservation issues, one would expect a growing political momentum as well as intellectual-academic constituencies being formed for achieving peace, security, equity and sustainability through effective ocean governance in the Indian Ocean.

Non-Traditional Threats to Maritime Security

Despite political, military, economic, social, cultural and language differences, varied national interests and threat perceptions, the countries of the Indian Ocean Rim need to cooperate on issues of common maritime security concerns, such as terrorism, piracy, gunrunning, disaster management, and search and rescue. As emphasized by Vijay Sakhuja in this collection, the networks of pirates and the terrorists have not only kept pace with the advancing technologies, including GPS and satellite communication systems, they have also proved to be highly skillful in adapting themselves to counter the strategies of maritime forces. It has been convincingly argued by P.V. Rao (2003:1) that the conventional state-centric and state-driven strategies of securing the seas, even though relevant in their own way, are in need of fundamental revision, in view of the fact that threats to maritime security also emanate from "a network of non-state actors, who are violent, illegal and organized". The fuzzy interconnectedness of such transnational threats to the maritime environment is graphically captured in the following quotation:

> The maritime element of the world economy should be seen as a global system characterized by countless interconnections such that a disturbance in one component will affect all the others. A hijacked oil tanker could founder on a distant reef, cause pollution, require naval forces to break off from their exercises, put up marine insurance rates . . . devastate local fisheries, set local countries at odds with each other and ruin a nearby tourist resort. This kind of example shows that the maritime economy has to be thought of and treated as a whole. (Till, 2001: S1)

However, the efforts made by several countries to address this problem have remained by and large bilateral, and, at best, trilateral in nature. What is needed, sooner than later, is a multilateral approach and greater cooperation among Indian Ocean regional states. De-territorialized threats of this kind seriously undermine the pursuit of economic growth and commercial interdependencies, and hence cannot be left entirely at the discretion or disposal of state actors and governments – a reality insufficiently acknowledged by the corporate world or for that matter NGOs. Combating transnational crime calls not only for high levels of intergovernmental cooperation and commitment; also essential is the creation of more economic opportunities, through international cooperation among state and non-state actors at both bilateral and multilateral levels, in the Indian Ocean's economically deprived sub-regions.

Environmental-Human Security

In the wake of the world-wide political and economic transformations of the 1980s and 1990s, increasingly complex discussions concerning North-South relations, and the growing focus on environmental issues, the language and metaphors of sustainability and sustainable development have spread very rapidly indeed. Few would deny that discourses structured around a range of issues such as biodiversity, climate change and ozone depletion, leading eventually to sea-level rise, carry some empirical substance. According to the Independent World Commission on the Oceans (1998: 98-99):

The problem that confronts us is that ocean ecosystems are often used in ways that are unsustainable, not only in environmental, but also in economic and social terms. All too often, the costs of this bear most heavily on the poor . . . An estimated 70% of the world's fish stocks are already being exploited at or beyond sustainable limits, but fishing generally continues unabated despite extensive regulatory arrangements for their management . . . The oceans have also become the ultimate sink of discharges of waste of all sorts – carried by rivers and winds – from land-based sources, including coastal megacities. Other threats come from the transport of hazardous wastes, operational and accidental spillage of oil, discharge of radioactive materials at sea, nuclear testing, and the transport of alien species in the ballast of ships.

While the problems, such as those above as examined by the World Commission on the Oceans no doubt exist, the specification and prioritization of ecological 'threats' and 'ocean governance' cannot simply be taken as scientific-technical-informational issues awaiting the attention of the governments or 'expert' intervention. According to the Commission, some of the major barriers to good ocean governance are: the proliferation of autonomous and semi-autonomous legal regimes and institutions, each with its own constituency among national government agencies, and net centrifugal tendencies in programming and financing; the scarcity of even basic ocean information in some countries, lack of public access to information, and fragmentation at the national level, and so on.

While few would question the relevance of the findings and recommendations of the World Commission, there is little evidence to suggest that the environmental processes and problems 'naturally' dictate an 'interested community' or that various communities and constituencies within states are affected by environmental degradation in the same way. A key challenge before the state as well as non-state actors, therefore, is to develop a normative vision of ocean governance and management, cognisant of the 'transparency' of the 'boundaries' both horizontal (as among disciplines, departments, ministries, specialized agencies and programmes) and vertical (as among levels of governance: local, national, regional and global), and committed to the principles of sustainable development, comprehensive security, equity, common heritage and participation.

The Independent World Commission on Oceans has also called upon states to cooperate and uphold international law and order at sea, while demonstrating much greater commitment to ecologically sustainable development and management of resources. A new approach to the question of equity in the oceans, to quote the Commission once again:

> . . . must give expression to the notion of human solidarity – a solidarity that engages all people, while being protective of differences based on outlook, region and culture. The approach must be guided by a sense of moral purpose and of long-term goals. It must be rooted in arrangements that serve the common good and responds to the interests and needs of different users of the oceans, incorporating the overarching imperative of sustainable development. In effect, the achievement of more equitable outcomes depends on the fashioning of win-win situations that reconcile different values and interests, and that rely on persuasion, mediation and compromise rather than on coercion, exclusive claims of right, and the domination of technological advantage. (Ibid.: SS)

As pointed out by Timothy Doyle in his thought-provoking contribution in this volume, neglect of environmental security issues will lead to increases in human conflict, and, ultimately, widescale disease, poverty and death. In view of their regionally-shared nature, these problems are best approached and tackled only through multilateral dialogue(s) based on a shared geopolitical vision and agenda for proactive action.

Sea Lane Security in the Indian Ocean

Yet another critical area which demands and deserves regional collaboration in the Indian Ocean is the safety of sea lanes of communication. While the major maritime trade route through the Indian Ocean is of increasing importance to the global economy, it funnels trade through vulnerable `strategic choke points' at the eastern and western exits as welt as the strait of Hormuz for traffic entering or exiting the Persian Gulf. Credible threats to those choke points could severely disrupt maritime trade, increase costs and impact on the health of the global economy. Freedom of the seas and particularly the free flow of commerce and maintaining the safety and security of the major maritime trade route that connects East Asia, Europe and East Coast are of critical (inter)national importance.

Threats to Sea Lane security in the IO have and would continue to come mainly from pirates, terrorists and other criminals sponsored by militia or insurgents. For reasons like traffic congestion, navigational constraints and the presence of economically deprived communities particularly along the coastal areas on the Indonesian side of the waterway, the Malacca and Singapore Straits are likely to remain a favourite venue for piracy activities. Concerns, bordering urgency, regarding the impact of sea robberies in the Straits on the economy and the potential risks to lives, properties as well as environmental damage, cannot be addressed through the installation of coastal maritime surveillance radar chains along the entire coast of the Straits and the deployment of more patrol boats alone. Long-term effective measures require regional and sub-regional agendas for dialogue as well as action.

What is also likely to make the SLOCs contested sites is the prospect of maritime rights opening up new areas of functional disagreement between maritime powers and smaller littoral states. Whereas the former are concerned with the threats of 'creeping jurisdictions' by littoral states on SLOCs accessibility, the latter are becoming increasingly anxious over 'creeping security jurisdiction' of the maritime powers especially over the high seas. However, the common stake of both the maritime powers and the littoral states in securing the SLOCs dictates the regional importance of their cooperation to ensure that security. Some of the compelling issue-areas that demand and deserve serious attention of both state and non-state actors are: SLOCs and environmental security; multiple uses, maritime law enforcement and the role of ports and coast guards; problems and prospects of regional and sub-regional cooperation; non-traditional threats to sea lane security: piracy/shipjacking, terrorism and political instability in littoral states; maritime trade among the littoral states of the IO; UNCLOS and freedom of navigation: legal regimes, maritime boundaries and geopolitical imperatives, and regional implementation of new security measures such as the ISPS Code and new seafarer ID documents.

Energy Security

It is a truism that the global demand for petroleum and natural gas continues to grow. According to the U.S. Department of Energy (2003), total world consumption of petroleum is expected to increase by 54 percent between 2001 and 2025, from 77.1 million barrels per day (mbd) to 118.8 mbd; the global demand for natural gas is projected to grow by 86 per cent over this period from 90.3 million cubic feet (tcf) in 2001 to 175.9 tcf in 2025. Although many producers will supply the petroleum needed to meet the import requirements of Asia and the United States, export production increasingly will be concentrated in a few major locations, notably the Middle East, the former Soviet Union and Africa.

The Indian Ocean and the states on its littoral are of growing importance. The region contains one third of the world's population and a quarter of its landmass. Currently, about 25 percent of all oil used by the United States passes through the sea lanes of the Indian Ocean and the Persian Gulf region, and the United States depends on the Indian Ocean for the shipment of about 50 different strategic materials, including tin, nickel, iron, lead, and copper. Two Indian Ocean states outside of the Persian Gulf, Burma and Bangladesh, which are rich in minerals, also are seen by key states as important future energy sources.

Easy, cheap and accessible oil for almost all industrialized and industrializing countries of the world can only be found in the Persian Gulf. All strategists have emphasized the enduring importance of the Persian Gulf after the withdrawal of the British from the region. Approximately two-thirds of the world's proven oil and one-third of its natural gas reserves are controlled by the Persian Gulf states. In case the estimated reserves of the Caspian Sea Basin are added to these figures, the respective percentage for reserves goes higher, perhaps 70 percent far oil and over 40 percent for natural gas. For this reason, as emphasized above, the Persian Gulf and the Caspian energy taken together are being perceived by the West as the most significant geo-strategic realities of the 21st century.

However, as Fereidun Fesharaki (1999) has pointed out, energy security concerns have shifted to Asia. High economic growth rates in Asia are likely to spur energy demands and serious conflicts might emerge as a result. Even though this demand suffered a temporary decline in the wake of Asian economic crisis, contributing to the slide in world oil prices and causing concern to producing countries, Asia is likely to remain at the forefront of the energy security debate. It is significant to note that spectacular growth in Asia has been followed by an equally spectacular growth in oil demand. Between 1990 and 1997, Asian oil demand grew by more than 6 million barrels per day (bld). During the same period, Asia accounted for 82 per cent of the global demand growth. In fact, Asia's oil demand overtook Western Europe's in 1990 and that of the United States in 1997. Within a decade, therefore, Asia became the largest oil-consuming region in the world (Ibid).

Indian Ocean Rim Association for Regional Cooperation (IOR-ARC): Realities, Perceptions and Policies

The fusion of India's "Look East" policy (against growing dissatisfaction with the painfully slow progress of the South Asian Association for Regional Cooperation) with Australia's "Look West" policy, combined with post-apartheid South Africa's need for a regional identity and the widely perceived need to respond effectively to the challenge of globalization, resulted in the fairly enthusiastic launch of the Indian Ocean Rim-Association for Regional Cooperation (IOR-ARC) in 1997. The IOR-ARC initially had 14 member countries (Australia, Indonesia, India, Sri Lanka, Singapore, Malaysia, Oman, Yemen, Kenya, Tanzania, South

Africa, Mozambique, Mauritius, and Madagascar). In 1999, this expanded to 19 countries (with Thailand, Bangladesh, Seychelles, United Arab Emirates and Iran). In addition, there are five dialogue partners (Egypt, Japan, China, UK and France) and one observer (Indian Ocean Tourism Organization). Australia, India and South Africa appear to be the three key players.

Although not explicitly stated in the IOR-ARC Charter, and despite arguments made both for and against, "open regionalism" was eventually adopted as the organization's guiding principle. (Essentially, open regionalism does not bind member countries to any commitments, is based on voluntary action, and decisions are taken on the basis of consensus.) This perhaps was not surprising, given the commitment of the leading economies in the region, including Australia, to the World Trade Organization.

However, as early as 1998, the IOR-ARC, seen by many initially as complementary to existing regional organizations such as APEC, the EU and NAFTA, had lost its direction as well as momentum. Such a crisis of disorientation appears to be have been caused by several factors and forces, especially the following:

- *First*, as a tripartite institutional arrangement (comprising business, academics and officials) imposed from above, without an established background of regional cooperation among either business or academic groups, the IOR-ARC appears to be lacking meaningful roots within the region. There is therefore a pressing need to cultivate both academic and business interest in the IOR-ARC, which at present is exercising minimal influence among member states.

- *Second*, the pro-active role and enthusiasm of Indian business organizations such as the Federation of Indian Chambers of Commerce and Industries (FICCI) and Confederation of Indian Industries (CII) and their counterparts in Oman has not been matched by the business interest from other countries, including Australia, Singapore and South Africa. An uncritical acceptance of the APEC model to begin with, combined with a relative neglect of realities specific to the Indian Ocean region – where bilateralism remains restricted mainly to certain sub-regions – has led to a situation where expectations are not matched by an overarching vision, political will, or the necessary building blocks.

- *Third*, "if IOR-ARC was looking for purely economic gains, then strategy should have been to bring more dynamic economies in the grouping in a very exclusive grouping from the very beginning. And emphasis should have been on economic integration as opposed to economic cooperation. In a loose grouping like IOR-ARC, concrete achievements are very difficult to demonstrate, and, in the absence of this, it is difficult to sustain the interest of policy-makers and business people in the grouping" (Sachdeva, 2003:6).

- *Fourth*, it was largely overlooked that, "trade and investment cannot be divorced from consideration of the environment; human resource development; maritime issues ranging from fisheries to shipping; communications; culture and security" (McPherson, 2002: 25). The lesson to be learnt here is that trade and investment are parts of a much more holistic view of ecologically sustainable, culturally sensitive and socially just development.

It is time to restructure and reorient the IOR-ARC along the lines of a *new* regionalism, dictated and driven by functional cooperation in sectors marked by growing convergence of interests among states and the corporate world and willing to adapt to new realities. As Baladas Ghoshal (2002) puts it:

For the time being, whatever sphere of cooperation the members may conceive for IOR, it has to be piecemeal and on the basis of a broad-based consensus. A collective action to prevent over-exploitation of the ocean's resources and over pollution may perhaps bear immediate results. Pollution of the Ocean has reached alarming proportions, the more so because of deep-sea fishing in exclusive economic zones (EEZ) of the littoral states. Eager to earn foreign exchange, these states are renting out their respective zones to foreign companies for large-scale fishing. Most of the catch is meant for export, but mechanized methods and tankers introduced for deep-sea fishing have disturbed the entire ecology of the Ocean. Coastal fish-breeding has nearly stopped, thus securing the livelihood of many fishing communities. This has forced them to migrate into the cities for alternative sources of livelihood or just vagrancy.

The current inward-looking, autonomy-seeking, geopolitically-driven, Euro-centric style of regionalism, so typical of the 1950s and 1960s, is likely to further reinforce a "govern-mentality" that is exclusively concerned with institution building. Instead, what the Indian Ocean demands and deserves, in our view, is an *adaptive* regionalism, which is outward-looking, and based on broadly conceived, flexible, market-sensitive, functional-sectoral cooperation. What is of critical importance here is not so much a commonly accepted *definition* of an Indian Ocean 'region', as an acknowledgement of the fact that the Indian Ocean is part of the world-ocean, and lacking inclusive networks of cooperation willing and able to *confront* uncertainty, *promote* confidence building and *incorporate* uncertainty.

In a recent thought-provoking article, Kenneth McPherson (2002) has argued that both the South Asian Association for Regional Cooperation (SAARC) and IOR-ARC need a new vision as well as agenda for regional and sub-regional cooperation, in order to achieve legitimacy, authority and effective governance. Such an agenda could possibly include far less contentious issues such as sea bed resources, fishing, illegal immigration by sea, transnational crime, maritime disaster management, meteorology, and search and rescue procedures. He further argues that, by joining IOR-ARC with an agreed agenda on maritime issues, SAARC would not only enhance its appeal as well as ability to further the process of developing regional cooperation, it could also contribute towards the rejuvenation of IOR-ARC. According to McPherson (2002: 260), "a shift in focus to include maritime issues may present an opportunity to break the logjam facing both organizations. They both need to establish new agendas to further the process of developing cooperation, and the Indian Ocean provides a range of vital issues that will profoundly affect all the region's inhabitants in the coming century."

In contrast to the cautious optimism of Kenneth McPherson, we have a skeptical view that, "there is nothing like an Indian Ocean region – at least not in any meaningful sense beyond mere geography – and that the prospects for security cooperation on the Indian Ocean as a whole are bleak" (Lehr, 2003: 1). The question posed by Peter Lehr is: "is it at all possible to identify parts of the Indian Ocean where the interests of policy-makers, of business people and epistemic communities do overlap, thus creating a common ˋmind map' conducive to multilateral security cooperation?" (Ibid.: 8). According to Lehr, the Milan East – social gathering of naval personnel – as a confidence building measure in the security sphere and the Bangladesh-India-Myanmar-Sri Lanka-Thailand Economic Cooperation (BIMST-EC) in the economic field are embedded in India's *Look East* policy. After the Bay of Bengal, argues Lehr, the Arabian Sea could turn out to be a "second possible spot for multilateral security cooperation which could be linked up with the Bay of Bengal" (Ibid.: 9). To quote Lehr:

> Take the case of Milan: Because of its success as a *confidence building measure*, India organized a similar venture in the Arabian Sea in the year 2000 as Milan West with the navies of Iran and Oman as participants. Again, India is a driving force, but not for altruistic reasons. Rather, Delhi has its own *energy security* in mind, which forms the back drop of its new Look West policy. Be that as it

may, *Delhi's security interests could ensure an initial take-off and a growth of more ambitious multilateral approaches – especially so since India alone cannot possibly safeguard the security of the Sea Lanes of Communication* (SLOCs). (Ibid.) (emphasis added)

A closer scrutiny would reveal that what we have chosen to call 'cautious optimism' of Kenneth McPherson is not as poles apart perhaps from the 'cautious pessimism' of Peter Lehr as might appear at first glance. We too had argued in the introductory chapter to this volume that the academic and the practical policy definition of the Indian Ocean as a 'region' is an inevitably contested issue. This should not come as a surprise because all regions are constructed, and, depending on the primary purpose of this construction – academic, administrative, economic, or, whatever – different regional definitions will follow. In short, the Indian Ocean, as a constructed region, will be defined by those who are involved in the construction process and whose collective goals will determine the composition of regional states. From the perspective of various types of regionalisms – economic, security and environmental – regional definition, of course, also depends on those states willing to join and remain with the group. The Indian Ocean region is no exception to the fact that the 'inclusive-exclusive' debate runs through all experiments in regionalization.

Having said that, what is of utmost importance for any long-term 'success' of any 'regional' or 'sub-regional' initiative in maritime issues – which cannot generally be treated as simple national issues – is the willingness to adapt to the boundless impulses and imperatives of flows through the strategy of networking and partnerships. Both McPherson and Lehr acknowledge the importance of an agenda-specific, bilateral/sub-regional sectoral-functional cooperation acting as a building block for wider 'regional' architecture. Consequently, the Indian Ocean is approached and understood as a 'Region of regions' – but always in the making. The percepts and decisions needed to transform such a vision into reality will not be like a pyramid with the apex sustained by the bottom, hegemonic and dominating, but influenced by the majesty of the oceanic circle, where the outermost circumference will not yield power to crush the inner circle but give strength to all within and will derive strength from it. Such a perspective opens up the space for SAARC joining IOR-ARC and the Bay of Bengal linking up with Arabian Sea. Accordingly, one need not rule out the possibility of the Gulf Cooperation Council (GCC) and the South African Development Community (SADC) joining heads and hands to realize shared visions and common objectives.

Concluding the Inconclusive

In a number of critical areas, ranging from environmental-energy security to sea lane security, as this book has shown, there are common stakes in the Indian Ocean but the dialogue(s) among the stakeholders has just begun. Sustaining the dialogue on an enduring basis continues to require a sincere and systematic multilateral, multi-level and multi-cultural effort in the direction of making politics more civil-society oriented, transcending narrow state-centric concerns. Equally crucial would be the willingness and ability of politics to take cognizance of, and accordingly respond to, the ecological dimensions of geography: that is to say, the finite capacity of the natural environment to endure and assimilate human-induced changes. It is only through extending the scope of maritime security dialogue on the Indian Ocean far beyond conventional state-centric preoccupations, to incorporate examination of questions relating to ecological, economic and cultural security of communities in the context of emerging global civil society, that a truly Indian Ocean vision can be generated and sustained. Needless to say perhaps, the success of such thinking on the ground (it may have to compete and even coexist with the old geopolitics of mastering the oceanic space) will also depend upon radical changes in the perceptions, policies and practices of those who speak and act on behalf of modern nation-states and the respective governments.

A key impediment in the way of an Afro-Asian dialogue on making the Indian Ocean a 'zone of peace', of course, is no doubt the revival of strategic rivalry in the Indian Ocean. According to Don Berlin (2002:26):

> The Indian Ocean region – after roughly a decade in the strategic wilderness – is again becoming an arena of rivalry among a number of key powers and littoral states. US forays have been the most audacious and insistent, but India has also been trying to strengthen its influence and bolstering its military posture. Beijing's most important connections with the region have been its links with Pakistan and Burma.
> An attribute of global political geography in the 20th century was the salience of northern Europe and northeast Asia. We do not yet understand the political geography of the 21st century. However, no region is likely to feature as prominently in that geography as the Indian Ocean due to its combination of oil, Islam, and the likely rise – and probably mutual rivalry – of both India and China. The Indian Ocean, often characterized as "the neglected ocean" will be so no longer.

Whereas it might sound highly utopian to expect such rivalries to disappear entirely from the Indian Ocean, the challenge is to ascribe new roles and missions to navies: not war-fighting but peace-building, not competitive but cooperative, in the service of monitoring and surveillance, law enforcement, research and rescue, disaster relief and humanitarian assistance. In other words, one of the key challenges before regional and sub-regional dialogues is precisely not to let such alternative visions disappear into oblivion, and persist with the agenda of maritime security broadly defined.

Regionalization is likely to flourish in the Indian Ocean only if we can establish close and cooperative relationship 'within' and 'between' the regions and sub-regions. As pointed out by V. Suryanarayan (2002: 270), ". . . exploitation of living and non-living resources, development of maritime communications; ship building and ship repair, weather forecasting; prevention of pollution and combating maritime terrorism – these tasks, which are the exclusive responsibilities of individual countries at present, can best be accomplished through regional cooperation." Ample scope for cooperation also exists by way of sharing development experiences and expertise, and also by the transfer of technology through collaboration in industrial, technological and the services sectors. There is a vast market base for potential cooperation in trade where cooperation in specific sectors can be highly beneficial.

Moreover, indications are that in the short period since IOR-ARC was established (in 1997) definitions of Indian Ocean regionalism have changed and will undoubtedly continue to change as the menu of common interests continues to expand. Both in the short and the long run, the IOR-ARC, through regional, sub-regional and sub-sub regional dialogues might also like to address itself to the challenge of (i) establishing schemes of international financing of global purposes at sea including charges for the use of global commons; that is, the area beyond the exclusive economic zones; such as sea lanes, high sea fishing areas, deep sea mining areas, other economic uses of international waters and tourism; (ii) working out people-oriented schemes of ocean space utilization and local utilization of ocean resources – including alternative energy resources – for meeting the needs of the coastal mega cities; (iii) promoting collaborative ventures in ocean research development within the Exclusive Economic Zones so that the revenue is generated for investment in the coastal mega cites; (iv) tapping renewable energy resources through ocean thermal energy conversion, tidal energy, wind energy, and salinity-difference for the local consumption of coastal mega-cities, and (v) working out adaptive strategies to deal with the problems of droughts and floods. In the long run, developing non-exploitative economic uses of the coastal areas, such as water-sports and eco-tourism could also prove to be immensely useful.

A "new" mapping of the Indian Ocean region by the political and economic elite is called for, which, in turn, demands political will as well as ability to question "land-centric" mindsets and engage far more actively with issues related to maritime security broadly defined. A long due realization should now be forthcoming that such issues straddle the traditional 'civilian-military' divide – where the responsibility for the former is assigned to the independent civilian Ministries/Departments and Agencies, and the latter to Ministry of Defence and armed forces – and, therefore, demands a new institutional landscape as well as higher levels of coordination.

Against the backdrop of growing regional militarization, the diffuse complexities of terrorism insecurities related to energy flows and environmental degradation, the idea of broadening the scope of the IOR-ARC as an Indian Ocean-centric architecture of maritime cooperation, based on adaptive regionalism, and achieved through both bilateral and sub-regional initiatives, deserves the critical attention of both state and non-state actors with a stake in the region. The new challenge calls for burden-sharing and joint responsibility among governments, the corporate sector, NGOs and academia to help defuse mutual suspicions, conflicts and rivalries.

Finally, dialogue, according to the Oxford dictionary, is conversation; a piece of written work in conversational form; a kind of composition; a conversational text or tract. There are no guarantees, therefore, that a dialogue will automatically improve a situation or relations. That noted, the dialogue has to be an ongoing, self-enriching, search for alternatives, which, in turn, will inevitably assist various regional and sub-regional initiatives of peace and cooperation with vision and substance.

References

Berlin, D. (2003), Paper Presented at the Indian Ocean Research Group (IORG) Inaugural Conference: "The Indian Ocean in a Globalizing World: Critical Perspectives on the 21st Century", ICSSR Complex, Panjab University, Chandigarh, India, 18-22 November 2002.

Berlin, D.L. (2002), "Indian Ocean Redux – Arms, Bases and Re-Emergence of Strategic Rivalries" *Journal of Indian Ocean Studies*, 10(1): 26.

Borgese, E.M. (1998), *The Oceanic Circle: Governing the Seas as a Global Resource.* Tokyo: United Nations Press: S0.

Chaturvedi, S. (2003), "Afro-Asian Oceanic Dialogue: Imperatives and Impediments", *Identity, Culture and Politics*, 3(2): 125-142.

Fesharaki, F. (1999), "Energy and the Asian Security Nexus", *Journal of International Affairs*, 53(1): 85-99.

Ghoshal, B. (2002), "What Makes Regionalism a Success? ASEAN and IOR-ARC Compared". Paper Presented at the Indian Ocean Research Group (IORG) Inaugural Conference: "The Indian Ocean in a Globalizing World: Critical Perspectives on the 21st Century", ICSSR Complex, Panjab University, Chandigarh, India, 18-22 November 2002.

Kelegama, S. (2002), "Indian Ocean Regionalism: Is There a Future?" *Economic and Political Weekly*, 22 June.

Lehr, P. (2003) "Prospects for Multilateral Security Cooperation in the Indian Ocean; A Skeptical View". Paper presented at the international conference on "India and the Emerging Geopolitics of the Indian Ocean Region", sponsored by Asia-Pacific Centre for Security Studies, Honolulu, Hawaii, 19-21 August.

McPherson, K. (2002), "SAARC and the Indian Ocean", *South Asian Survey*, 9(2): 251-262.

Rao, P.V. (2003), "Indian Ocean: Non-Military Threats to Maritime Security". Paper presented at Conference on "Narratives of the Sea: Encapsulating the Indian Ocean World", organized by Nehru Memorial Museum and Library, New Delhi, 10-12 December.

Report of the Independent World Commission on the Oceans. 1998, *The Ocean: Our Future*. Cambridge: Cambridge University Press.

Sachdeva, G. (2002), "IOR-ARC at Crossroads". Paper Presented at the Indian Ocean Research Group (IORG) Inaugural Conference: "The Indian Ocean in a Globalizing World: Critical Perspectives on the 21st Century", ICSSR Complex, Panjab University, Chandigarh, India, 18-22 November 2002.

Singh, K.R. (2002), "Geo-strategy of Commercial Energy", *International Studies*, 39(3): 259-288.

Suryanarayan, V. (2003), "Plea for a New Regional Organization" *South Asian Survey*, 9(2): 263274.

Till, G. (2001), "A Changing Focus for the Protection of Shipping", *Journal of Indian Ocean Studies*, 9(1).

U.S. Department of Energy, *International Energy Outlook 2003*, Tables A4, AS.

Index

Aden, 41, 52
Afghanistan, 13, 77, 79
Africa, 92
  land degradation, 161
African National Congress (ANC),
  138
African Union, 140
Age of discoveries, 57, 79
al-Qaida, 71, 273, 275, 278
American naval power, 45
Andaman Sea, 250
Anglo-American relations, 79
Anglo-Japanese alliance, 44
Apartheid, 63, 72, 78, 80, 138
  Gandhi's struggle against, 78
Arab-Israeli conflict, 162
Arab states, 80
"Arc of crisis, 27, 56, 69, 79
Arc of militarization, 91
Archipelagos, 66, 244
Arid regions, 182
Arms trade, 14
ASEAN, 21, 27, 114, 127, 131, 139
Asia, 20
Asia-Pacific Economic Cooperation
  (APEC), 22, 114
Asia-Pacific region, 130
Asian Development Bank, 185
Asian Highway (AH)', 126, 127
Asian tigers, 61
Atlantic Ocean, 3
Australia, 1, 80, 86
  consumption patterns in, 158

defence policy, 8
fertility levels, 227
immigrants category, 200
India-born population, 222, 223,
  225
international migration issues, 226
'look west' policy, 24, 148, 306
migration from India, 200
multicultural society, 199
population, 220
population census, 197, 202
security of, 9
settlement patterns, 224
student arrivals, 213
terrorism, 12
Australia India relations, 17
Australia-Iran trade, 4
Australia-Sri Lanka trade, 4

Bab el Mandeb, 71, 287
Balance of power, 35, 74, 78
Bali, bombings in, 2
Bangladesh, 27
Bay of Bengal, 36
Bilateralism, 144
Biodiversity, 168
Black Africa, 63
Biodiversity, 302
Brain drain, 79, 218, 228, 233
British empire, 34, 65
British imperialists, 75, 179
British lake, 35, 42, 85
British naval power, 42